P9-ARO-503

NATURAL ORDER:
Historical Studies of Scientific Culture

SAGE FOCUS EDITIONS

1. **POLICE AND SOCIETY**
   edited by DAVID H. BAYLEY
2. **WHY NATIONS ACT: THEORETICAL PERSPECTIVES FOR COMPARATIVE FOREIGN POLICY STUDIES**
   edited by MAURICE A. EAST, STEPHEN A. SALMORE, and CHARLES F. HERMANN
3. **EVALUATION RESEARCH METHODS: A BASIC GUIDE**
   edited by LEONARD RUTMAN
4. **SOCIAL SCIENTISTS AS ADVOCATES: VIEWS FROM THE APPLIED DISCIPLINES**
   edited by GEORGE H. WEBER and GEORGE J. McCALL
5. **DYNAMICS OF GROUP DECISIONS**
   edited by HERMANN BRANDSTATTER, JAMES H. DAVIS, and HEINZ SCHULER
6. **NATURAL ORDER: HISTORICAL STUDIES OF SCIENTIFIC CULTURE**
   edited by BARRY BARNES and STEVEN SHAPIN

NATURAL ORDER

Historical Studies of Scientific Culture

edited by

Barry Barnes and
Steven Shapin

Beverly Hills/London

For information address:

SAGE PUBLICATIONS, INC.
275 South Beverly Drive
Beverly Hills, California 90212

SAGE PUBLICATIONS LTD
28 Banner Street
London ECIY 8QE England

Printed in the United States of America

Library of Congress Cataloging in Publication Data
Main entry under title:

Natural order.

(Sage focus editions; 6)
Includes bibliographical references.
1. Science—History. I. Title: Scientific culture. II. Barnes, Barry. III. Shapin, Steven. IV. Series.
Q125.N36 509 78-19650
ISBN 0-8039-0958-6
ISBN 0-8039-0959-4 pbk.

FIRST PRINTING

CONTENTS

ACKNOWLEDGMENTS

The editors are grateful for the technical assistance of Ann Baxter and Moyra Forrest in preparing this volume for publication. They should also like to express their appreciation of encouragement and critical comment received from colleagues in the Science Studies Unit, Edinburgh University, and elsewhere. Particular debts are owed to David Bloor and Martin Rudwick.

INTRODUCTION

Perhaps the most signifcant change in the history of science, and indeed in the study of science generally, over the last decade is that it has become more relaxed and naturalistic. Increasingly, we have become prepared to treat science as an aspect of our culture like any other. The intense concern of earlier generations with the special status of science and its allegedly distinctive characteristics has begun to ebb away. Without any great proclamation or catastrophic upheaval in method, scholars have been increasingly willing to accept accounts of scientific change simply as the techniques of their discipline reveal it to them, and to perceive less and less need for a "rational reconstruction" of the past. There is now a real interest in our natural knowledge as a product of our way of life, as something we have constructed rather than something which has been, so to speak, revealed to us.

If this tendency continues, if we relax completely about natural knowledge and set it alongside technique, art, music, or literature as simply a part of our culture, then some interesting consequences will follow. One is that the contrast of "internal" and "external" factors in the history of science will cease to be a major source of interest or controversy. That external procedures, representations, or even standards of judgment become taken up into the esoteric subculture of "science" will simply be something to be noted and treated as seems fit to the interpretive task at hand. External factors would then be dealt with as they presently are when taken up into the subcultures of painters or musicians. Nobody expects a subculture to be capable of complete independence of its environment or general setting; nor, in the case of most subcultures, do historians condemn their productions if autonomy is found to be imperfect. Another consequence is that those who study natural knowledge will feel free to experiment with any of the general methods and theories of the social sciences. Insofar as such methods and theories appear to have merit in the context of art or religion, or the cosmologies of preliterate societies, or any other setting, they may prove useful in the study of science also. As a typical form of culture, science should be amenable to whatever methods advance our understanding of culture generally.

It might appear at first that what is being talked of here is a possible reversion to the state of affairs before the last world war when a significant movement of Marxists and radicals sought to *expose* science as a function of its social context. But this is not at all the case. In the 1930s both sides of the great debates accepted the importance of the internal/external dichotomy, and both sides recognized that sustaining their discourse were opposed methods of evaluating science and opposed policies towards it. In contrast, the current move to naturalism is not systematically related to any distinctive evaluation of science or policy towards it. Naturalism closes no evaluative or political options; it merely ejects them from historical practice. It may be, indeed, that the main impetus to naturalism stems from professional, disciplinary considerations on the part of historians and others engaged in the study of science. A recognition that explicit evaluative concerns and commitments are not conducive to good history may be the major factor. If one aspires to *do* history in a properly "disinterested" way, it is difficult simultaneously to act as an apologist for science, making out its past as a disembodied interaction between rational minds and reality.

Thus, if the trend we are talking about continues (which is far from certain), the movement is likely to be from the celebration and sacralization of science to more descriptive and analytical approaches, and not to its defamation. If a philosopher for the trend is required, then the "conservative" Wittgenstein will serve as well as Marx, and, given current tastes in the history of science, is likely to be preferred. Wittgenstein insistently points out that all our classifications are conventions, that we have made them all, that it is our practice which sustains them—or changes them. Thus, he helps us to maintain real curiosity about the whole of our culture and to remain aware of its total contingency: he dissuades us from giving up the search for understanding with lazy references to reality, or nature, or logic, or necessity. But nowhere does Wittgenstein equate the contingent or conventional with the worthless, the socially sustained with the groundless; and nowhere does he treat our own activity as a subversion of what it has, after all, produced.

Unfortunately, these very associations remain all too common today, even among practising historians of science. Doubtless, they serve to sustain the spurious distinction between the internal intellectual history of the rational growth of knowledge, and the external social history of science, dealing with alleged irrational influences upon it. Probably also they support the current hierarchy on which the two approaches are situated. No doubt these associations will continue to be invoked in lieu of argument by those philosophers who still insist that sociological and anthropological theories cannot possibly have relevance to the study of rational thought, and who frequently compound their ignorance with egregious arrogance in the enunciation of their dictates. As historians of science know very well, the secularization of an area of enquiry takes time to accomplish.

The best way to establish the possibility of doing something is to do it. That science stands symmetrically with other forms of culture can readily be shown in terms of abstract arguments, which are abundant and are now widely known. But if the point is really to be given recognition, it needs to be demonstrable by pointing to cases wherein it is evident. This too is perfectly possible at the present time, but a combination of circumstances makes existing work of this kind less visible and recognizable than might be expected. In the social sciences, where explicitly theoretical work on forms of culture is routinely accepted, there has been, until very recently, remarkably little interest in the knowledge and practice of the natural sciences, and competent analyses of aspects of scientific culture are not yet widely disseminated. Indeed, in the United States, where unlike Europe there has been an active interest in science for a long period, it was tacitly accepted by sociologists that the theories of their discipline should not be applied to scientific knowledge per se. In history, on the other hand, there has been a number of important concrete studies of the social basis of scientific knowledge and even of its intimate connection with the general social context. But other factors than those operating in sociology have kept these achievements from being recognized for what they are. First, they have been widely dispersed in the literature; a collection or anthology of such studies has never been produced. Hence, new approaches have diffused thinly into the existing framework of the history of science, and, to an extent, have been interpreted by and assimilated into existing patterns of thought. Second, methods of training historians of science have not, in general, provided them with the social science competences which would facilitate such new perspectives on science, even though many social scientists are themselves hostile to a sociological interpretation of natural knowledge. And finally the individualism and particularism of much of the most celebrated history of science have provided concrete exemplars of what sorts of practice are most likely to be accepted in the academic community. As explicit theorizing is routinely discouraged in history of science, the import of existing achievements in the historical sociology of scientific knowledge has been difficult for many to perceive. In certain instances, even the authors of such studies deny the significance of their own work.

Thus, it occurred to us that there would be reason for a book to stand as a visible symbol of this new tendency in the history of science. We have sought to compile a collection of original essays *explicitly* concerned with natural knowledge as culture, and *explicitly* exploring the possibilities of anthropological and sociological methods in understanding it. There have indeed been collections of essays in the history of science wherein the possibility of sociological approaches to science has been debated, or programmatically discussed. But, as we write, we know of no other collection wherein the feasibility of the approach is simply assumed, and the aim has become that of implementing it, and showing how various kinds of analysis which fall within its ambit can be applied in particular cases.

We asked our contributors to be as deliberately exploratory and speculative as they liked, to raise questions and to exhibit the possibilities of various kinds of approach. We recognized that such a style would run counter to the normal idiom of history of science, and that it could be regarded as unduly theoretical. But we have no apology whatever to make for this. Theoretical and reflective thinking, generalizing and comparative historical work, and even rank speculation, all have their roles to play in historical practice: they need no defence. They have their contribution to make as much as the most meticulously detailed particular studies. However, a collection of essays which were *solely* programmatic and speculative in the worst sense would rightly be regarded with scepticism. Even to illustrate and explore the possibilities of new methods, it is best if they can be shown at work, getting to grips with particular materials. Hence, for the most part, we have invited contributors to write on the basis of detailed empirical studies which they have already carried out or are presently carrying out. Although, for reasons of space, full documentation and normal detailed exposition of material is not possible here, the essays are typically connected with material extensively documented elsewhere, which may be referred to at need.

We are grateful to our authors for being willing to embark upon the hazards of this atypical enterprise, for agreeing to be more reflective than they might normally be, and for leaving much of their documentation elsewhere. The reader is asked to blame the editors rather than the authors for these indiscretions, and to accept that in these particular essays all the evidence for the cases being made is not to be found within the text itself. To take these essays as documented demonstrations of specific hypotheses, fully worked-out and self-sufficient, would be to miss their point. But to take them as suggestions for what *might* be done would be equally misconceived: in most cases they present what *has* been done, concretely, in proper detail, but they do so under the constraint of restricted space, and with unusually explicit stress on general themes and theories.

Our predominant concern has indeed been to obtain contributions properly based in concrete work, and for this reason no unified point of view, or overall framework or theory, will be found consistently used and advocated through the book. We have solicited contributions with a naturalistic approach, or with an interest in anthropological and sociological theories of culture, but we have not tried to impose any further unity upon them. Disagreements and clashes of emphasis will thus be found from paper to paper. We have, however, grouped the essays into loosely thematically related sets, in introducing which we give a brief indication of the content of the various particular contributions and their bearing on themes of general interest.

Barry Barnes
Steven Shapin

Part One BODY ORDER

Any perceived pattern or organized system in nature is liable to be employed to express and comment upon social order and social experience. In being so employed, the perceived pattern is itself liable to be developed and reconstituted to better fit it to its functions. The pattern in question may be of many kinds: the overall order of the cosmos, the system of natural kinds of plants and animals, the general organization of the earth's crust, even the humdrum routines of the honeybee.

One of the most notable achievements of social anthropology, and particularly of the school of Emile Durkheim, has been an elucidation of the character of this relationship between social and natural order in preliterate society. This kind of research remains significant in modern social anthropology, notably in the work of Mary Douglas, where it is extended from the restricted context of preliterate cultures and turned into a completely general approach—a potential resource for understanding our own natural order as set out by scientific practice. It is encouraging to find that the detailed theories of Professor Douglas are now beginning to be taken up, used and tested by historians and sociologists of science.

Of all the models of coherent organization recognized in nature, none is more readily available than the human body, and none is more frequently turned to as a tool for explaining or justifying social order. Thus it is not surprising to find that recent work by historians of medicine and the life sciences indicates that an anthropological orientation to Western anatomy and physiology can be very fruitful. The three essays in this section offer some findings and illustrate a little of what research can hope to reveal within this area. None of the three papers is primarily concerned to demonstrate *that* social experience and interests affect conceptions of the body; rather, each *proceeds from* this realization to argue for some more specific link between "the body natural" and "the body politick." Lawrence's and Shapin's papers are broadly compatible, both in the context they examine and in the orientations they employ. In the mid-18th century we see Scottish Lowland elite groups figuring their social situation in an indigenous representation of bodily function. Their small-group solidarity is mirrored by their view of the human body as integrated and controlled by the nervous system and "sensibility." In turn, the order of the body, once accredited, is available to them as a justification of their social dominance. Shapin deals with the same Edinburgh context more than a half-century later on. Now the hegemony of old elites is being threatened by emergent bourgeois groupings. Old representations of the relationship between mind and brain, between man and nature, between the individual and society, are being challenged by new, more materialistic, mechanistic and naturalistic attitudes. Phrenology is the opening move in the campaign of 19th-century scientific naturalism discussed in Part Two. Shapin's approach is designed to show how the theoretical perspectives which social anthropologists have developed in the light of their fieldwork can be profitably brought to bear in the interpretation and explanation of detailed historical materials.

If the arguments of these two essays are accepted, then they establish two important general points of continuing relevance throughout the book. First, both essays point out that conceptions of the natural order may be multifunctional. Both identify social interests bearing upon the production of representations of the body, but both also emphasize that the "technical" value of those representations should not be thought of as thereby diminished. Social influences are factors to be described, and their consequences delineated; they are not factors to be "exposed" for their corrupting effects on scientific knowledge. Secondly, it is a mistake to imagine representations of natural order being, as it were, *constructed* by an examination and direct rendition of reality, and then being *used* in a social context. Representations are not first constructed, and then assessed, and then used. As historians, all that we encounter is use. Representations are constituted and reconstituted, assessed and reassessed, continually, in the process of use. Accordingly, they cannot be studied by methods which assign them independent, inherent characteristics (of meaning, or implication, or truth) prior to their use.

Many of the themes of Lawrence's and Shapin's papers are also found in Roger Cooter's theoretical discussion of early 19th-century popular physiology and its reception. But in one respect it offers a significant contrast. Lawrence and Shapin both attempt to interpret and explain theories and representations by showing how they were used and by whom. They propose to understand ideas by referring them to people, their activities, and their interests. In Cooter's discussion, on the other hand, the power of ideas and ideology over people is also asserted, albeit with the qualification that the potency of ideology is strictly dependent on the socioeconomic context wherein ideology is deployed. Given proper conditions, he suggests, knowledge can 'grab people' against their own interests. Cooter says that artisans accepted popularized physiology as a manifestation of reason and rationality, and "Failing to perceive the ideological power that Reason has assumed. . .became its victims, destined. . .to promote and safeguard the Reasonable bourgeois world." Cooter's approach is frequently adopted by European Marxist intellectuals, and is not otherwise represented in this volume. Doubtless, some may find it curious to encounter Marxist interpretations stressing the historical potency of ideas and ideologies in opposition to what it is tempting to call the "bourgeois materialism" of writers such as Shapin. But this kind of stress has, in fact, always featured as a significant strand of Marxist theory.

1

THE NERVOUS SYSTEM AND SOCIETY IN THE SCOTTISH ENLIGHTENMENT

Christopher Lawrence

It has long been recognized that many features of Scottish science in the 18th and 19th centuries distinguished it from science elsewhere. Recently, scholars have turned their attention to the causes of this singularity, in particular to tracing the relations between science and other areas of culture and to the relations between science and aspects of the Scottish social structure.[1] Medicine in Scotland, which at the beginning of the 18th century was broadly similar to medicine elsewhere in Europe, had, by mid-century, acquired distinctive characteristics. Morrell (1971) has pointed to some of the institutional reasons for its success and style. In physiological theory Scottish medicine was characterized by its stress on the total integration of body function, the perceptive capacity or sensibility of the organism, and a preoccupation with the nervous system as the structural basis for these properties.

AUTHOR'S NOTE: For their helpful advice and criticism I should like to thank Barry Barnes, Michael Neve and Sallie Rée. In this respect I owe a particular debt to Steven Shapin. I am grateful to John Christie and Francis Doherty for allowing me to examine unpublished work. For permission to quote from manuscripts I am grateful to the Keeper of the Records of Scotland. For their financial support of this research I should like to thank the Wellcome Trust.

All these concerns can be found in contemporary European physiology, but nowhere were they combined and stressed in the same way. This *uniqueness* of Scottish medical theory is, I suggest, only explicable by invoking the particular context of the upheaval in Scottish economic and social life in the 18th century, and, in particular, by referring to the social interests and self-perceptions of the improving landed class that came to dominate Scottish culture. The concept of the body in the cosmology of this class, and the physicians who articulated it, furthered more than a single interest. Through a theory of sensibility, physiology served to sanction the introduction of new economic and associated cultural forms by identifying the landed minority as the custodians of civilization, and therefore the natural governors, in a backward society. A related theory of sympathy expressed and moulded their social solidarity. And it goes almost without saying that, by ordering and explaining physiological phenomena, such concepts satisfied physicians' technical demands.

A CIVILIZED SOCIETY

The general form of the social and economic history of the Scottish Enlightenment is now becoming clear. Though debate rages as to which factors were of most significance, a framework has been established within which a new interpretation can be given of that brilliant effulgence of culture which, to choose only a few disparate areas of activity, produced such men as Adam Smith, David Hume, a pair of Allan Ramsays and Joseph Black. A potentially fruitful account of the social background to this intellectual efflorescence has been given by Phillipson (1973, 1975). The incorporating Union with England of 1707 was a turning point in Scottish history, though whether for good or ill continues to divide historians (Trevor-Roper, 1977). The following 75 years divide into three roughly equal periods. The first, until the mid-1720s, was a time of economic depression and social fragmentation. From 1720 to 1750 the economy began to revive and the future culture of Edinburgh began to take shape. The period 1750 to 1780 witnessed the first fruits of the economic boom and the blossoming of intellectual life. After this time political and religious tensions reemerged and a commercial middle class made its voice felt in both economic affairs and natural philosophy (Shapin, 1979).

The years prior to the Union were ones of economic failure and cultural impoverishment. In those immediately following, the ruling oligarchy, left as a disorganized rump after the loss of their parliament, cast around for strategies which would give them at once social legitimacy as a group and also form the basis for cultural and economic growth. By mid-century this strategy had crystallized out as a campaign for the *improvement* of Scottish life, a campaign which served the economic ambitions of the ruling elite and the cultural pretensions of a new city literati. This philosophy of improvement can be regarded as ideological in that it was designed to further the landed class's interests in achieving social legitimacy and authority.[2]

By the 1720s the old oligarchy had turned away from the commercial schemes of the 1690s and had begun to organize themselves around a policy of agricultural improvement; by 1723, they had formed a Society for that purpose. The economic success of Scottish agricultural change in the 18th century is well known (Handley, 1953; Hamilton, 1963). From being one of the most backward countries in Europe, lowland Scotland became an agricultural trend-setter. Ancient farm towns gave way to enclosed lands; new crops were experimented with; a mobile labour force was created. Other land-based economic enterprises also flourished: coal mining, sheep farming, forestry, and the establishment of planned villages and estates. Following Pocock (1965), Christie (1975:117) has argued that "Improvement partook strongly of neo-Harringtonian ideology"—that is, it emphasized the desirability of the landed gentry exercising paternalistic authority over tenantry, as well as over legal and ecclesiastical life. That authority is what the lowland landowners came to achieve: "As landlords they had the reputation of being the most absolute in Britain" (Smout, 1969:282).

The landowners' campaign for agricultural improvement had its counterpart in the programme of cultural improvement that burgeoned among the intellectuals of the capital. In the 1720s the University was remodelled around a new medical school (Christie, 1974; Morrell, 1974, 1976). Among the city literati debate centred on social, intellectual and moral improvement and took place in the multiple clubs and societies that were formed in the city. One important facet of this improving ideology was the cultivation of manners and polite literature. Scottish intellectuals turned to English forms in poetry, prose (especially the gentile variety of Addison and Steele), language (Hume, it is said, died regretting his Scotticisms—not his sins), and social habits. Refinement, delicacy, and moderation were the key synonyms for culture by mid-century (Smout, 1969; Plant, 1952; Graham, 1969).

In the full flowering of the Edinburgh Enlightenment in mid-century the improving gentry adopted the cultural forms and patronized the activities of the city's literati. The latter then, in turn, were sanctioned as the articulators of a specifically *Scottish* cultural identity. The fusion of interests had begun informally in the 1720s; many landowners were lawyers and inevitably belonged to both groups. By the 1750s the shared ideology of improvement had developed institutional forms. Most notable was the very select Select Society in which philosophers, doctors, and aristocracy discussed and articulated the ideals of economic and cultural progress (McElroy, 1952). The landowners also patronized and shaped the activities of natural philosophers, coupling a fashionable interest in science with a belief in its value for promoting an improved agriculture (Shapin, 1974). Physicians inevitably were key figures among the city's cultural and social leaders. One other feature confirmed the integration of mid-century society, the newly emerged "polite Church" (Smout, 1969: 231). The dominant Moderate Party turned away from the austere Calvinism of their fathers and pursued the rational theology of the Enlightenment. They wrote histories, debated in societies, and eventually attended the theatre. Improvement, integration and custodianship of the values of landed society were the Church's policy (Chitnis, 1976:43-74; Drummond and Bulloch, 1972).

The members of the ruling caste which emerged in mid 18th-century Edinburgh, and which gradually consolidated their power over Scottish economic and cultural life, were obviously not without their similarities; they were identifiable both by background and by interests. They were largely Presbyterian landowners, often judges or advocates, Whigs with a commitment to the Union and an allegiance to the London government. The University, for instance, was a Whig stronghold (Horn, 1967:36-94). Family ties, as well as identity of cultural interests, often served to align them with the city literati and churchmen who spoke the same "language." The preoccupation of this group with its solidarity is well recognized (Chitnis, 1976:36-40), and this preoccupation was more than idle gossip. There were disruptive elements in Scottish life striving for power and independence over whom the lowland elite strove to exert economic and social control: the fanatical church, the new commercial classes, and, most importantly, the highlands, where economic control and political subjugation were essential to a thriving lowland agrarian economy.

The seat of Jacobitism, Catholicism, the home of a society based on the clan system and speaking a foreign language, the highlands were regarded by the lowlanders as the seat of barbarianism and a potential source of insurrection. The policy of the London government to the highlands was, by turns, one of foresight and bungling ineptitude, concern and indifference. By the time of the 1745 Rebellion the government had come to need a sledge hammer to crack the Jacobite nut (Mitchison, 1970). The stringent statutes following Culloden served to consolidate the power of the Whig clans, especially the Argylls whose formidable power controlled most of the patronage in Edinburgh. The aftermath of the Rebellion also served to strengthen the elitism of Edinburgh society. But the Rebellion was not in itself a crucial turning point in Scottish life; in many ways it served only to accelerate changes initiated in the lowlands which had been going on for some time previously. Before the Rebellion, under the policy of the House of Argyll, that change had begun which resulted by the end of the century in the economy of the highlands being changed so radically "that its near independence was transformed into almost total subordination to the demands of lowland industry" (Cregeen, 1970:9). Gradually at first, and increasingly after the Rebellion, economic pressure induced the highland chiefs to succumb to lowland policy and transform their land use. Such pressure, however, resulted not only in an economic change but in the adoption of lowland forms of dress, manners, religion, speech, and political sympathy.

The assimilation of the highlands was pursued by the lowlanders within the structure of their ideology of improvement which contrasted savage highland society with civilized values. This ideology served to identify the nature of a socially divisive force, exert pressure to change on the cultural forms in which, apparently, rested its military and anarchic character, and legitimate the controlling ambitions of the lowland elite. The assimilation of the highlands to lowland control in the 18th century was the consequence of a complex blend of a raw economic pressure and the new more 'civilized' cultural forms that both sanctioned and activated it.

THE PHYSIOLOGY OF CIVILIZED MAN

The Edinburgh Medical School was founded in 1726. The drive for its foundation came from a small number of men in the city who were highly sensitive to the town's economic decline after the Union. They saw in the revitalization of the University, and especially in a new Medical School, both a way of retaining in the city the flood of Scottish students who went to Holland, and a means of attracting foreign students to the town (Christie, 1974; Morrell, 1974, 1976). The new medical school was modelled on the Dutch one at Leyden which its greatest professor Hermann Boerhaave had fashioned to his own particular ends. The improvement of the medical faculty proved a sound project. By mid-century it was the most successful in Europe, both in terms of student numbers and in the acclaim of its teachers. The professors themselves became some of the most famous and influential men in Edinburgh; they had great power in the University, most of them having been appointed through the patronage of the Argylls; they belonged to the most exclusive clubs; they were all Whigs and moderates in religion; some were landowners and of course improvers.

The first five professors of the new Medical School, of whom the most famous is Alexander Monro *primus,* had all studied at Leyden in the years following the Union. The medical curriculum they taught was identical to Boerhaave's both in form and in content, and it dominated Edinburgh medicine until mid-century. In Boerhaave's physiology the body was essentially a complicated hydraulic machine, the concept being the direct intellectual descendant of the Cartesian model. The account placed little emphasis on the overall coordination or integration of body functions and was concerned primarily with the dynamics of the blood vascular system. Man in this model has a second substance, the soul, the repository of sensations and will, unconnected with the vital functions (Lawrence, 1976).

By the mid-18th century the nervous system had come to dominate Edinburgh physiology. The major teachers of this period were William Cullen, Robert Whytt, Alexander Monro *secundus,* and John Gregory. Although the nervous system gradually assumed dominance in the whole of 18th-century physiology, in Edinburgh this dominance took a very particular and interesting form. In a recent paper Figlio (1975) has pointed to some of the reasons for the 18th-century preoccupation with neural function. Most important of these were the Lockean epistemology and psychology with their basis in sensation, the growth of experimental neurology, and a general shift in 18th-century thought from reductionism to crypto-vitalism, or, as Schofield (1970) has said, from mechanism to materialism (cf. Brown, 1974). Also of great importance was a new interest in the environment as a determinant of man's nature and of civilization generally (Glacken, 1969:551-622; Porter, 1977).

The physiological correlate of this environmentalism was the development of the concept of the reactive organism. Given a sensualist epistemology, the nervous system was clearly going to be of importance as the mediator between man and his environment. Inevitably, then, on the physical condition of the nervous system would depend the quality of sensation (Jordanova, 1977). The upshot of these conceptual shifts was a move from Cartesian dualism to monism, with the nervous system itself as the bridge which possessed attributes of both mind and body. In consequence certain key terms such as "sensibility", and "susceptibility to impressions" were used interchangeably as definitions of the properties of the nervous system or of the soul. The sensibility of the nerves, of course, varied with their physical state; disease, for instance, could change sensibility. In consequence, the quality of the sensation experienced by the soul varied, and certain descriptions such as "coarse," "delicate," "refined," and so on, applied both to the state of the nervous system and to the quality of the received perceptions. Given a sensualist psychology, then, all man's higher attributes—taste, imagination, and, indeed, the capacity to reason—would, in the last analysis, depend on his conditions of existence, diet, weather, labour, and so forth. The theory of sensibility acquired an unusual importance in Scotland because Scottish philosophy posited *feelings* as the basis of all human actions.

The first intrusion of the nervous system at an influential level in Edinburgh medicine occurred with the creation of Robert Whytt as Professor of Medicine in 1747. Whytt came from a landed legal family and was a Whig whose appointment had been approved by Argyll (French, 1969:1-16). Tolerant in religious matters, he was sufficiently clubbable to be a manager of the Select Society (McElroy, 1952:149). Though an Edinburgh alumnus, he rejected the doctrine of a "mere inanimate machine" and by 1739 had reintroduced the soul into the body (Whytt, 1768:1; French, 1969:5). Whytt's principal contention was that the body's responses are purposeful and not the result of blind mechanism. A blink, for example, is a deliberate response to threatened injury and not a concatenation of pressure effects. The source of the body's purposeful behaviour was what he termed "the sentient principle"—an immaterial, undivided substance that could "feel" stimuli and necessarily directed the appropriate response. Only in some cases did the feeling reach consciousness—for example, in the irritation of the bladder preceding micturition. In other cases, such as the heart beat, the stimulus was "felt" unconsciously (French, 1969:71; Whytt, 1768:98). The sentient principle, being a feeling agent, was seated by Whytt in the nervous system, with which it was coextensive. The principle, and thus the system, became the seat of overall control and integration of body function.

Whytt brought the nervous system to Edinburgh and with it the death sentence for Boerhaavean physiology. Among all the important teachers of the second generation, the nervous system was the focus of interest and in many ways their teachings are no more than extensions and variations of Whytt's doctrines. William Cullen came to Edinburgh in 1755 and was probably the most successful teacher the Medical School had. He was worshipped by his students and his doctrine, more then any other, became the new medical orthodoxy. Cullen was the personification of Enlightenment Edinburgh. He was involved in numerous schemes concerning the application of chemistry to manufactures and agriculture, and in later life owned an estate himself (Thomson, 1859:ii, 673). He belonged to the most exclusive societies and was an intimate of Hume and Smith; his religious and philosophical views were clearly formed partly under their influence (Donovan, 1975:34-76).

Cullen's teachings differed from those of Whytt in that, wanting physiology to be an autonomous science, he was not prepared to posit the existence of immaterial sentient principles outwith the bounds of natural law. Neither did he want a physiology reducible to mechanical principles. The outcome was a model of the body based on an excited state of an aetherial fluid in the nervous system. This nervous power was unique to life and may be "properly enough termed the 'vital principle'" (Cullen, 1789:i, 59). Like Whytt's sentient principle, the nervous energy is the repository of sensibility and, like it, is excited "By certain internal impressions . . . [which] . . . are the causes of the motions of the heart and arteries" (Cullen, 1827:i, 110). Cullen, in other words, retained all the characteristics of Whytt's sentient principle—purposeful action, coordinating ability, and, most importantly, unconscious feeling—without introducing second substances into physiology.

The other two central figures in Edinburgh medicine at this period were Alexander Monro *secundus* and John Gregory. Monro, like Cullen, was an important contributor to 18th-century Edinburgh cultural life, both as a teacher and as a member of society at large. He, too, was a close friend of Hume and Smith. Not least was he preoccupied with agricultural innovation on his estate (Wright-St. Clair, 1964:92). John Gregory was in some ways an outsider. He came to the Chair of Medicine in 1766 from Aberdeen, where, with his cousin Thomas Reid, he helped develop the philosophy of common sense, the doctrines of which recurrently appear in his writings (Olson, 1975). Gregory moved in the usual elite Edinburgh literary and philosophical circles, but always remained slightly shy of Hume (Carlyle, 1860:461). Godly common sense was, after all, at war with scepticism.

Both Gregory and Monro taught a concept of the body that incorporated the thinking of Whytt and Cullen. Both affirmed the existence of a sentient principle, but postulated an electric nervous energy *as well*. For both of them the sentient principle coordinated bodily activity by sensibility. As Monro (1783:86) pointed out, there are two kinds of feeling "one with and another without consciousness," or as Gregory said, it "depends greatly on the mind and yet is without consciousness" (1769:66).

Edinburgh physicians developed a model of the body in which sensibility, a property of the nervous system, predominated. Using this concept, they were able to provide a physiological and anatomical basis for one of their primary concerns, overall integration of the body functioning. This they did by using the notion of "sympathy"—which was no more than the communication of feeling between different bodily organs, manifested by functional disturbance of one organ when another was stimulated. Sympathy was thus a special case of sensibility, and dependent on the same factors. The doctrine of sympathy, conceived as the mutual feeling between different parts, had first been formulated by Whytt in about 1744 (Whytt, 1768:241)[3]. The term itself has a long history. Before Whytt the prevailing confused opinion was that sympathy was mediated by the intercostal nerve described by Willis (1681:157). Others thought sympathy was transmitted by the blood or tissue contiguity (French, 1969:33). Whytt was the first to give the term clearly defined structural and functional significance (though he was "wrong" to ignore the importance of the intercostal nerve).

Sympathetic action for Whytt is always centrally mediated through the nervous system (1768:492-524). Using the concept, Whytt did his most brilliant experimental work, notably on the effect of opium on the nervous system, on the inhibitory nature of shock, and his description of what is now termed reflex action and its limitation to part of the cord (French, 1969:63-92). This concept of sympathy was endorsed by all the other leading Edinburgh teachers. Cullen praised Whytt's work (Thomson, 1859:i, 258), though he was a little unhappy about the term sympathy since he thought it was redolent of occult qualities (Thomson, 1859:i, 306). Monro *secundus* went so far as to plagiarize entirely Whytt's account of sympathy and claim it as his own (1783:100)[4]. Gregory showed a similar concern with bodily coordination by sympathy, "an association of feelings communicated . . . by nerves" (1769:66).

Sympathy, though generally the unconscious physiological coordinator, could also reach consciousness; for instance, Whytt noted that "by doleful stories or shocking sights delicate people have often been affected with fainting and general convulsions" (1768:493).

Such an effect is clearly on an epistemological continuum with somatic sympathy, for it is only the transmission of feeling by the nerves:

> although in these cases the changes produced in the body are owing to the passions of the mind; yet the mind is only affected through the intervention of the optic and auditory nerves, they seem proper enough instances of the general sympathy that extends through the whole nervous system. [Whytt, 1768:493]

Gregory gave a similar example: "people subject to hysteric fits will often by sympathy fall into a fit by seeing another person fall into the same" (1769:66).

Sensibility and its special case, sympathy, were thus the foundations of Edinburgh physiology. Since these were fundamental properties of the nervous system, identification of the factors affecting them was central to any theory of normal function and the basis of a concept of disease. Sensibility, Cullen held, could alter in intensity or *quality* (the latter he regarded as his own discovery). He classified the anatomical, physiological and psychological elements affecting sensibility as (1) the anatomy of the sense organ (the particular thickness of tissue between stimulus and nerve), (2) heredity, (3) vascular influences, (4) temperature, (5) previous impressions, (6) state of the nerves, (7) the state of brain, (8) degree of attention (1827:42). These elements, given the environmentalist bent of 18th-century physiology, were held, in turn, to depend predominantly on such factors as diet, exercise, climate, and so on—features traditionally known as the "non-naturals," which though not always so called, were the basis of most 18th-century aetiological theory (Rather, 1968; Jewson, 1974). Sensibility, then, in the end related to the individual's mode of life and should, in the healthy state, be properly adjusted to it. Change of sensibility ouside an individual's normal state produced disease. This concept was first clearly elaborated by Whytt in his important work on disease of the nerves (1764), and became in principle the conceptual basis of nearly all mid-century Edinburgh pathological theory. As Cullen said, "the nerves are more or less concerned in every disease" (Thomson, 1859:i, 343).

THE SENSIBILITY OF CIVILIZED MAN

The nervous system gained importance in Scotland not only from the physiological side of the bridge but from the philosophical side also. In the second quarter of the century Scottish philosophy turned from reason to feeling, both as the basis of morals, and as the mainspring of action. On this foundation was developed a sophisticated theory of society and history, as well as a school of sentimental novel-writing (Bryson, 1945; Thompson, 1931).

The major writers of mid-18th-century Edinburgh were, of course, David Hume and Adam Smith. Their works are significant not only because of posterity's judgment, but because their contemporaries judged them "important" and deemed them "influential" (Phillipson, 1973). However, the themes discussed below are not to be found solely in their works, but also in those of very many mid-18th-century Edinburgh literati. Scottish social thinkers discerned a relation between social life and the quality of sensibility. History was a process involving a gradual refinement of feeling. Rude uncivilized peoples and the labouring poor were hardly sensible—a condition both consequent on, and necessary for, an arduous physical life. Insensibility referred, indistinguishably, to physical or social feeling. Smith noted that the hard life, coarse diet, and continuous danger endured by the savage numbs him to the feelings of others and to physical pain: "He can expect from his countryman no sympathy or indulgence . . . Before we can feel much for others we must in some measure be at ease ourselves" (1976:205). Likewise, the savage can be "hung by the shoulders over a slow fire" with apparent indifference, whilst "the spectators express the same insensibility" (1976:206). Lord Kames, who referred approvingly to Smith's moral theory (1774:ii, 312), made the same association between the different sorts of feelings:

> a savage enured to acts of cruelty feels little pain or aversion in putting an enemy to death . . . but . . . a person of so delicate feelings, as scarce to endure a common operation of phlebotomy . . . will be shocked to the highest degree, if he see an enemy put to death in cold blood. [1751:145]

The same insensibility is found to a lesser extent in the European counterpart of the primitive, the labouring poor. As Hume pointed out, "the skin, pores, muscles and nerves of a day-labourer are different from those of a man of quality so are his sentiments, actions and manners" (1974:ii, pt.3, 115). The medical writers provided the physiological foundation for this account. Cullen, for instance, in his relation of the factors influencing the sensibility of the nervous system, specifically noted the "difference between inhabitants of a rude and uncultivated climate and those of a polished and civilised nation" (1827:i, 40). Gregory, discussing the nervous temperament characterized by excessive sensibility, noted that "this temperament commonly attends the rich, indolent and luxuriant. The labouring people are never attacked with nervous disease" (1769b:219). The dependence of the mental life on the properties of the nervous system was also made explicit by the moral philosophers, albeit metaphorically. Adam Ferguson (1966:92) noted, it is

> in admiration of fortitude . . . that the Americans are so attentive, in the earliest years to harden their nerves. The children are taught to vie with each other in bearing the sharpest torments.

With the development of increased sensibility, there arose the refined, delicate feeling that was the bond of the true society and the guarantor of taste.[5] It was necessarily the preserve of the few who had time and opportunity to cultivate their imaginations. As Smith said, "the amiable virtue of humanity requires surely, a sensibility much beyond what is possessed by the rude vulgar of mankind" (1976:25). Or, as Lord Kames put it, "sensible people endeavour to correct an appetite . . . and it is . . . the vulgar chiefly who allow themselves to blindly to be led by the present instinct" (1751:29).[6] Hume likewise was concerned to demonstrate that fine discernment was based on sentiment and that not all judgments were equal. A delicate aesthetic sense, though not the same as a physical feeling, has as its precondition a subtle perception:

> a man in a fever would not insist on his palate as able to decide concerning flavours; nor would one affected with the jaundice pretend to give a verdict with regard to colours. In each creature there is a sound and defective state; and the former alone can be supposed to afford us a true standard of taste and sentiment. [1760:ii, 373]

Refined taste, as Hume made clear, develops from the shared sentiments of the elite few:

> though men of delicate taste be rare, they are easily to be distinguished in society by the soundness of their understanding, and the superiority of their faculties above the rest of mankind. [1760: ii, 387]

Gregory, the physician, endorsed this:

> the advantages derived to mankind from taste by which we understand the improvement of the powers of the Imagination are confined to a very small number. The servile condition of the bulk of mankind requires constant labour for their daily subsistence. This of necessity deprives them of the means of improving the powers either of imagination or reason. [1772:104][7]

This theory of refined sensibility with its naturalistic basis in the nervous system and its corollary in a theory of the true custodians of civilized society, identified the natural governing role of the Edinburgh improvers, and sanctioned the cultural changes they were introducing. It differentiated them from the population at large and from recalcitrant landowners addicted to uncouth and boorish habits and old-fashioned agricultural techniques, but most of all from the savage highlander. Hume's attitude to the "barbarous" highlanders and their "obstinate adherence to their ancient customs" is typical (1970:142). It is not difficult to extract from *The Wealth of Nations* a picture of a starved and backward example of humanity. Highland life was hard, the climate cold, the diet meagre, coarse, and monotonous. The physiological correlate was hard, indelicate nerves lacking in sensibility. The cultural effect was the martial, superstitious life beyond the highland line.[8] This ideology of sensibility clearly served to identify the economic backwardness and cultural depravity of highland life and place it in a cosmology with a strong progressive historical dimension.

THE SYMPATHETIC SOCIETY

The Scottish theory of man's actions rested on feeling. Besides this the philosophers developed a theory of social bonding, and in some cases ethics, based on a special case of feeling, sympathy. This approach to social theory had originated with Francis Hutcheson, Professor of Moral Philosophy at Glasgow from 1730 to 1740 (Davie, 1952, 1961, 1965, 1973). His attempt to develop a view of society different from the individualistic vision of Hobbes was renewed in the second part of Hume's *A Treatise of Human Nature* of 1739. Besides the passions and sentiments, Hume identified a further important constituent of human nature, sympathy, which is not a passion but

> [T]hat propensity we have to sympathise with others, to receive by communication their inclinations and sentiments however different from or even contrary to our own. [1974:ii, 40]

Sympathy is a transmission of sentiment, when we feel it "we feel a sensation" (1974:ii, 99). On the basis of this doctrine Hume built his theory of society, for it is sympathy that makes man what he is, a social creature (Stewart, 1963). With this theory Hume had moved away from the 18th-century fascination with the individual's invariable characteristics and began to develop a theory of society as something more than the sum total of its individual components (Wertz, 1975; Morrow, 1923). The principle of sympathy was taken over from Hume by Adam Smith and used by him as the basis for his social and ethical theory (Campbell, 1971). Smith's account of sympathy is found in his *Theory of the Moral Sentiments* of 1759, though he was clearly teaching it in the early 1750s (Smith, 1976:Introduction). Sympathy, for Smith as for Hume, is a feeling. As such it is transmitted by the nervous system and, therefore, depends on the latter's condition. Smith makes this quite explicit in his first example of sympathy:

> Persons of delicate fibres and a weak constitution of body complain, that in looking on the sores and ulcers which are exposed by beggars in the streets they are apt to feel an itching or uneasy sensation in the corresponding parts of their own bodies. [1976:10]

Such statements might equally well be found in contemporary medical texts. Smith, of course, was generally more concerned with the transmission of emotions, but the principle is identical; the basis for the quality of these feelings is the delicacy or otherwise of the nervous system. Sympathy, and, therefore, social bonding, increases with the progress of society; as a form of sensibility, it is therefore most highly developed in the minority free of the hardships of everyday life. Like gravity (and the analogy was evidently in Smith's mind), sympathy decreases with distance; in other words, it is greatest among small groups of men, characterized by division of labour, but joined by a common purpose. (This incidentally has been proposed as the solution to "*das Adam Smith Problem*"—the "contradiction" between the sympathy of the "early Smith" and the apparent individualism of *The Wealth of Nations* (Wills, 1978).) Since sympathy is the basis of Smith's ethical theory, as sensibility is the basis of taste, it is clear from his analysis that anything worthy of the name of civilization is carried by a small group bound together by strong mutual feeling. Smith could indeed conceive of a society based on utility alone—one of merchants, for instance—but he noted it would be "Much less happy and agreeable than one in which assistance is reciprocally afforded from beneficent natures" (1976:85). The implicit defence of landed paternalism against mercantile interests is clear.

The preoccupation with solidarity based on sympathy is not unique to Hume and Smith. It is found, for instance, among the Common-Sense philosophers (Davie, 1973). The physician and philosopher John Gregory, for example, wrote that "The next distinguishing principle of mankind . . . is that which unites them into societies and attaches them to one another by sympathy and affection" (1772:86). Alexander Monro *primus*, on the other hand,who belonged to the first generation of physicians and who also wrote an essay on society made no mention of sympathy (1739). Neither did sympathy receive attention in his medical works. Among the other philosophers, Lord Kames (1751:17) called sympathy "the great cement of human society." A similar concern with "sympathy, humanity, fellow feeling

[and] social solidarity" can be found in Adam Ferguson (Forbes, in Ferguson, 1966:xxvii). The same is also true for writers on law (Stein, 1970) and moderate churchmen (Chitnis, 1976:59). Hugh Blair, the most popular Presbyterian sermonizer, thought of sensibility as a sort of social glue without which, "men would become hords [*sic*] of savages" (1795:iii, 28). Thus, there is a strong case for regarding the 18th-century Scottish vision of society as unique (Swingewood, 1970).

In his (1967) essay on the Scottish Enlightenment, Trevor-Roper addressed the problem of why it was Scotsmen particularly who were interested in the social behaviour of mankind. He found the answer in the social circumstances of the thinkers, in the contrast they saw between the new world of rapid movement and their own hinterland of antique custom. It was a perceptive insight that begged elaboration. Preoccupation with social integration developed at a time when, through an ideology of improvement, city literati and improving landlords sought successfully to further their common interests. The daily proof of their intimacy of contact was the myriad projects, clubs, societies, and the like, that flourished in the city. Here there existed a fluidity of social exchange among a section of Scottish society that remained unparalleled elsewhere; this apparent conviviality and sense of integration were to become a source of great nostalgia at the beginning of the 19th century (Shapin, 1975). Outside of that closed social nexus were the barbarous highlands, unimproved areas of the lowlands, fanatical religious sects, a nascent mercantile class, and a labouring population recalcitrant to agricultural and industrial change. Scottish philosophers indeed perceived a difference between the backward countryside and the mannered life of the town. The observation, however, was not entirely disinterested.

Concern with solidarity extended not only into social theory but also into physiology. As society was held to depend on the mutual feeling, or sympathy, between the parts, so, too, was the body. In the same way that the Edinburgh elite perceived itself, so the nervous system was seen as a structure of interacting sensibilities, binding together and controlling the whole. There was no room for autonomous functions in the description of the body and society. Nor, indeed, was there in Scotland itself. Rather than looking, then, as is so often done, for the "influences" of philosophy on medicine or the reverse, both must be interpreted by referring them to the common social context. To do this is not to suggest the model of the body was a *mere* celebration of the social order. It arose concurrently with that social order and, as part of the whole ideology of improvement, served to give the social leaders their strong sense of identity and to sanction their natural governing role. But this was not all; instrumentally it was at the same time a brilliantly innovative exploration and interpretation of physiological evidence.

POSSIBILITIES

There is not space here to sketch more than a few possible relations of this account to a more orthodox history of 18th-century physiology, which I see being extended rather than supplanted by the present orientation. Conventional historiography would, I think, see the shift in Edinburgh physiology from mechanism to neural integration as being intelligible in terms of two major factors: first, a general European shift from mechanism to vitalism; second, an increasing interest in the nervous system following from the studies of Willis in the 17th century and Whytt himself in Edinburgh. This sort of account has already been supplemented by Brown (1974), who sees in the dominance of the varieties of mechanism in England up to the 1730s the endeavour of the London Royal College of Physicians to align itself with the power elite in politics and the church. Similarly, it is possible that in Scotland the adoption of Boerhaave's Newtonianism meant alignment with Whiggism, the moderate wing of the Kirk, and the ideals of the Union (cf. Jacob, 1976). Such an explanation would not, of course, conflict with the economic model of Morrell (1974, 1976) and Christie (1974). Schofield (1970) has suggested that, well before mid-century, mechanism was failing, as a philosophy of nature, to 'deliver the goods' in explaining chemical and physiological phenomena. This may be so, but is also seems clear (Brown, 1974) that Scottish physicians were turning to Continental vitalist theory as a counter to the hegemony of the London College of Physicians from which they were excluded (Waddington, 1973). There was, it should be stressed, nothing to be gained in therapeutic efficiency in the shift from mechanism to vitalism. Clinical practice remained almost identical under the new theory.

This chapter's emphasis upon the social context was intended to help comprehend the distinct form and local meaning of Edinburgh physiology. Specifically, it was designed to explicate the concatenation of vitalistic theory, stress on neural integration and control, mental and social philosophy, and social legitimation which was found nowhere else. In other contexts, vitalism assumed very different forms: in London John Hunter ascribed his "living principle" to the blood, and sometimes to the solids as well (1794); in France the major thinkers developed a vitalism stressing the *independent* sensibility of each organ (Haigh, 1976); in Göttingen Albrecht von Haller (1752) elaborated his theory of "irritability" as an autonomous property of muscle. In each "foreign" case, the model of the body was more decentralized than that developed in Edinburgh. It was precisely this decentralization that Edinburgh theorists rejected, subsuming all functions under the control of the nervous system. The fate of the integrative model later in the century should prove a fertile area for study, especially when examined in relation to changing physiological theory and social circumstances. It is clear, for instance, that certain pupils of William Cullen, such as George Fordyce (1768, cf. 1791) and Gilbert Blane, who originally followed their teacher's doctrines, later on distributed the vital properties throughout the body. A further possibility for research might be the examination of the development of Brunonianism in 1770s Edinburgh as a reaction to University medical orthodoxy.

In conclusion, I should like to summarize my central argument in such a way as to relate it to other studies in this volume (such as those by MacKenzie and Barnes, Porter, Shapin and Wynne). In my attempt to elucidate the contextual significance of Edinburgh physiology I have discerned certain formal similarities between conceptions of nature (in this case, the body) and conceptions of society. The Edinburgh theory of the body and the Edinburgh theory of social order used a common concept, integration through feeling. I have sought, however, to say more about this similarity in physiological and social theory than that both deployed common material. I have tried to identify the specific social groups which elaborated social and natural thought, to ascertain their historical situation and social interests, and have attempted in turn to show that the model of the body was developed, evaluated, and deployed in a context of legitimation.[9] Thus, I would claim that physiological conceptions were sustained by social interests, and that the present account is broadly compatible with anthropological orientations to natural knowledge.

There is, however, a possibility that such a claim may be misunderstood. I have been at pains to emphasize that the Edinburgh model of the body was "multi-functional." Physiology in Edingburgh was undoubtedly influenced and moulded by its social context, just as it was undoubtedly influenced and moulded by "technical" interests in explaining natural phenomena. "Social influences" on a body of knowledge do not necessarily vitiate its power as a scientific tool. Using the Edinburgh model, Whytt and Monro *secundus* did a great deal of brilliant physiological and anatomical work and Cullen developed a whole new theory of fever. If one liked, one could argue that the modern theory of neural integration derives from this 18th-century Edinburgh work. If, then, the relation between social interests and scientific theory described above is correct, it serves to add to our historical understanding of a piece of science, just as sociologically conceived history of science serves to supplement, not to supplant, traditionally conceived "internal" history.

NOTES

1. This is not entirely true, for in a different tradition Buckle (1857) grappled with the distinctive nature of Scottish science. Of the more recent work Christie's essay (1975) provides a most suggestive framework for interpreting Scottish science. His first footnote is a guide to the best modern literature.

2. My use of the overworked term ideological is closest to that of Barnes (1977:27-44).

3. This paper had first been read in Edinburgh about 1745 (French, 1969:9).

4. From internal evidence I think it unlikely the concept was Monro's, most of whose life was plagued by priority disputes.

5. It is widely recognized that the Scottish writers, and especially Hume, placed little emphasis on the environment proper (climate, geography) as an important determinant of the nature of society. Rather they were more concerned to identify the moral and economic causes of social change. I am not disputing this interpretation. Given that society *is made by men's actions,* some individuals will inevitably have a poorer diet, be more exposed to cold, will need to labour harder, and so on. They will in consequence be equipped with the appropriate physiology. A change in society may result from economic or moral causes, but these in turn will affect the material conditions under which men live.

6. Another common theme developed by Scottish philosophers and medical writers was to regard animals, savages, and the labouring poor as governed mainly by instincts and the upper classes by sensibility.

7. A survey of Gregory's writings would demonstrate that he was often at pains to eulogize the noble savage. This, however, was coupled with an attack on luxury, and ergo on commercial society. It is quite clear that he preferred civilization to the savage state.

8. Attitudes to highlands habits were at times ambivalent, as indicated by the Ossian controversy. As intellectuals fluctuated between anglophilia and Scottish identity, the highlands could either be despised as barbarous and rude, or praised as the home of simple, unadorned, and natural society.

9. In practice, physiology was not often resorted to as a legitimation of social interests. Improvers were after all a powerful group and had only occasional need to buttress their arguments with naturalistic sanctions.

REFERENCES

BARNES, B. (1977) Interests and the Growth of Knowledge. London: Routledge & Kegan Paul.

BLAIR, H. (1795) Sermons (9th ed., 5 vols.). London.

BROWN, T.M. (1974) "From mechanism to vitalism in eighteenth-century English physiology." Journal of the History of Biology, 7:179-216.

BRYSON, G. T. (1945) Man and Society: The Scottish Inquiry of the Eighteenth Century. Princeton; N.J.: University Press.

BUCKLE, H.T. (1857) History of Civilization in England. London.

CAMPBELL, T.D. (1971) Adam Smith's Science of Morals. London: George Allen & Unwin.

CARLYLE, A. (1860) Autobiography. Edinburgh.

CHITNIS, A.C. (1976) The Scottish Enlightenment.London: Croom Helm.

CHRISTIE, J.R.R. (1975) "The rise and fall of Scottish science." Pp. 111-126 in M. Crosland (ed.), The Emergence of Science in Western Europe. London: Macmillan.

_____ (1974) "The origins and development of the Scottish scientific community 1680-1760." History of Science, 12:122-141.

CREGEEN, E. (1970) "The changing role of the House of Argyll in the Scottish Highlands." Pp. 5-23 in N.T. Phillipson and R. Mitchison (eds.), Scotland in the Age of Improvement. Edinburgh: University Press.

CULLEN, W. (1827)The Works (2 vols.). Edinburgh.

_____ (1789) A Treatise of the Materia Medica (2 vols.). Edinburgh.

DAVIE, G.E. (1973) The Social Significance of the Scottish Philosophy of Common Sense. Dow Lecture, Dundee University.

_____ (1965) "Berkeley's impact on Scottish philosophers." Philosophy, 40:222-234.

_____ (1961) The Democratic Intellect. Edinburgh: University Press.

_____ (1952) "Hume and the origins of the Common Sense School." Revue Internationale de Philosophie, 6:213-221.

DONOVAN, A. (1975) Philosophical Chemistry in the Scottish Enlightenment. Edinburgh: University Press.

DRUMMOND, A. and BULLOCH, J. (1973) The Scottish Church 1688-1843. Edinburgh: Saint Andrew Press.

FERGUSON, A. (1966) An Essay on the History of Civil Society 1767. Edinburgh: University Press.

FIGLIO, K.M. (1975) "Theories of perception and the physiology of mind in the late eighteenth century." History of Science, 13:177-212.

FORDYCE, G. (1791) A Treatise on the Digestion of Food. London.

_____ (1768) Elements of the Practice of Physic. London.

FRENCH, R.K. (1969) Robert Whytt, the Soul, and Medicine. London: Wellcome Institute.

GLACKEN, C.J. (1969) Traces on the Rhodian Shore. Berkeley: University of California Press.

GRAHAM, H.G. (1969) The Social Life of Scotland in the Eighteenth Century (5th ed.). London: A. & C. Black.

GREGORY, J. (1772) A Comparative View of the State and Faculties of Man compared with the Animal World (5th ed.). London.

——— (1769a) "Lectures on the institutions of medicine." Scottish Record Office, Acc. GD119/449.

——— (1769b) "Lectures on the institutions of medicine." Scottish Record Office, Acc. GD119/448.

HAIGH, E.L. (1976) "Vitalism, the soul, and sensibility: the physiology of Théophile Bordeu." Journal of the History of Medicine and Allied Sciences, 31:30-41.

HALLER, A. VON (1752) A Dissertation on the Sensible and Irritable Parts of Animals. First English edition 1755, reprinted in Bulletin of the Institute of the History of Medicine (vol.4.) (1939).

HAMILTON, H. (1963) An Economic History of Scotland in the Eighteenth Century. Oxford: Clarendon Press.

HANDLEY, J.E. (1953) Scottish Farming in the Eighteenth Century. London: Faber & Faber.

HUME, D. (1974) A Treatise of Human Nature. London: Everyman.

——— (1970) The History of Great Britain: The Reigns of James I and Charles I. Harmondsworth: Penguin.

——— (1760) Essays and Treatises on Several Subjects. London.

HUNTER, J. (1794) A Treatise on the Blood, Inflammation and Gun-shot Wounds. London.

JACOB, M.C. (1976) The Newtonians and the English Revolution 1689-1720. Hassocks: Harvester Press.

JEWSON, N.D. (1974) "Medical knowledge and the patronage system in eighteenth-century England." Sociology, 8:369-385.

JORDANOVA, L.J. (1977) "Views of the Environment 1750-1850." Paper presented to a conference on New Perspectives in the History of Geology, New Hall, Cambridge, April 1977; to appear in revised form as "Earth science and environmental medicine: the synthesis of the late Enlightenment," in L.J. Jordanova and R. Porter (eds.), Images of the Earth: Essays in the History of the Evironmental Sciences. British Society for the History of Science Monographs.

KAMES, H. HOME, LORD, (1774) Sketches of the History of Man. Edinburgh.

——— (1751) Essays on the Principles of Morality and Natural Religion. Edinburgh.

LAWRENCE, C.J. (1976) "Early Edinburgh medicine: theory and practice." Pp. 81-94 in R.G.W. Anderson and A.D.C. Simpson (eds.), The Early Years of the Edinburgh Medical School. Edinburgh: Royal Scottish Museum.

McELROY, D.D. (1952) The Literary Clubs and Societies of Eighteenth Century Scotland. Edinburgh University. Ph.D. thesis.

MITCHISON, R. (1970) "The government and the Highlands, 1767-1745." Pp. 24-46 in N.T. Phillipson and R. Mitchison (eds.), Scotland in the Age of Improvement Edinburgh: University Press.

MONRO, A. *primus* (1739) "An essay on the female conduct contained in letters from a father to his daughter." National Library of Scotland, Acc. 6659.

MONRO, A. *secundus* (1783) Observations on the Structure and Functions of the Nervous System. Edinburgh.

MORRELL, J.B. (1976) "The Edinburgh Town Council and its University." Pp. 46-65 in R.G.W. Anderson and A.D.C. Simpson (eds.), The Early Years of the Edinburgh Medical School. Edinburgh: Royal Scottish Museum.

_____ (1974) "Reflections on the history of Scottish science." History of Science, 12:81-94.

_____ (1971) "The University of Edinburgh in the late eighteenth century: its scientific eminence and academic structure." Isis, 62:158-171.

MORROW, G.R. (1923) "The significance of the doctrine of sympathy in Hume and Adam Smith." Philosophical Review, 32:60-78.

OLSON, R. (1975) Scottish Philosophy and British Physics 1750-1880. Princeton, N.J.: University Press.

PHILLIPSON, N.T. (1975) "Culture and society in the eighteenth-century province: the case of Edinburgh and the Scottish Enlightenment." Pp. 407-448 in L. Stone (ed.), The University in Society (vol. 2). Princeton, N.J.: University Press.

_____ (1973) "Towards a definition of the Scottish Enlightenment." Pp. 125-147 in P. Fritz and D. Williams (eds.), City and Society in the Eighteenth Century. Toronto: Hakkert.

PLANT, M. (1952) The Domestic Life of Scotland in the Eighteenth Century. Edinburgh: University Press.

POCOCK, J. (1965) "Machiavelli, Harrington, and English political ideologies in the eighteenth century." William and Mary Quarterly, 3rd series, 22:589-583.

PORTER, R. (1977) The Making of Geology. Cambridge: University Press.

RATHER, L.J. (1968) "'The six things non-natural': a note on the origins of a doctrine and a phrase." Clio Medica, 3:337-347.

SCHOFIELD, R.E. (1970) Mechanism and Materialism: British Natural Philosophy in an Age of Reason. Princeton, N.J.: University Press.

SHAPIN, S. (1979) "Homo phrenologicus." In this volume.

_____ (1975) "Phrenological knowledge and the social structure of early nineteenth-century Edinburgh." Annals of Science, 32:219-243.

_____ (1974) "The audience for science in eighteenth-century Edinburgh." History of Science, 12:95-121.

SMITH, A. (1976) The Theory of Moral Sentiments. Oxford: University Press.

SMOUT, T.C. (1969) A History of the Scottish People 1560-1830. London: Collins.

STEIN, P. (1970) "Law and society in eighteenth-century Scottish thought." Pp. 148-168 in N.T. Phillipson and R. Mitchison (eds.), Scotland in the Age of Improvement. Edinburgh: University Press.

STEWART, J. B. (1963) The Moral and Political Philosophy of David Hume. New York: Columbia University Press.

SWINGEWOOD, A. (1970) "Origins of sociology: the case of the Scottish Enlightenment." British Journal of Sociology, 21:164-180.

THOMPSON, H.W. (1931) A Scottish Man of Feeling. London: Oxford University Press.

THOMSON, J. (1859) An Account of the Life, Lectures and Writings of William Cullen MD. Edinburgh.

TREVOR-ROPER, H. (1977) "Covenant and union." Times Literary Supplement, September 9, Pp. 1066-1067.

_____ (1967) "The Scottish Enlightenment." Pp. 1635-1658 in T. Besterman (ed.), Studies on Voltaire and the Eighteenth Century (vol. 58)., Genéva: Institute et Musée Voltaire.

WADDINGTON, I. (1973) "The struggle to reform the Royal College of Physicians 1767-1771: a sociological analysis." Medical History, 17:107-126.
WERTZ, S.K. (1975)"Hume, history and human nature." Journal of the History of Ideas, 36:481-496.
WHYTT, R. (1768)The Works. Edinburgh.
——— (1764) Observations on the nature, causes and cure of those Diseases which are commonly called Nervous, Hyperchondriac or Hysteric; to which are prefixed some remarks on the sympathy of nerves. Edinburgh.
WILLIS, T. (1681) The Anatomy of the Brain. London.
WILLS, G. (1978) "Benevolent Adam Smith." New York Review of Books, February 9, Pp. 40-43.
WRIGHT-ST.CLAIR, R.E. (1964) Doctors Monro. London: Wellcome Medical Library.

2

HOMO PHRENOLOGICUS: ANTHROPOLOGICAL PERSPECTIVES ON AN HISTORICAL PROBLEM

Steven Shapin

In recent years increasing numbers of historians and sociologists of science have been concerned with identifying links between natural knowledge and social context. This represents a considerable shift from past traditions, which referred the generation of scientific knowledge to its "correspondence" with reality, and the institutionalization of science to the recognition of its manifest truthfulness. The first steps towards a contextual history and sociology of science have, therefore, already been taken. The number of concrete studies which analyse the production, evaluation, and acceptance of scientific knowledge as social processes is large and growing larger. The continued assertion that scientific knowledge is autonomous and transcendent is now more often founded on ignorance of this new literature than on considered criticism thereof (Hammerton, 1977; cf.

AUTHOR'S NOTE: I should like to express my gratitude to the following for critical comments on an earlier version of this paper: Dr. R.H. Barnes of the Institute of Social Anthropology, Oxford; Barry Barnes and David Bloor of the Science Studies Unit, Edinburgh; Professor Yehuda Elkana of the Van Leer Foundation, Jerusalem; Dr. C. J. Lawrence of the Wellcome Institute for the History of Medicine; and Dr. Roy Porter of Churchill College, Cambridge.

Durant, 1977). Traditional formulations of the "internalist-externalist" debate are now superseded (MacLeod, 1977). A new task confronts the historian of science.

The task is the refinement and clarification of the *ways* in which scientific knowledge is to be referred to the various contextual factors and interests which produce it. The mere assertion that scientific knowledge "has to do" with the social order or that it is "not autonomous" is no longer interesting. We must now specify how, precisely, to treat scientific culture as social product. We need to ascertain the exact nature of the links between accounts of natural reality and the social order. In short, we need to provide a social epistemology appropriate for the history of science (Barnes, 1974, 1977; Bloor, 1976; Barnes and Bloor, forthcoming).[1]

Empirical studies which *explicitly* address themselves to this project scarcely exist in the history of science. However, the history of science literature is replete with examples which may be studied with this interest in mind. The problem here is that a richness of explanatory techniques and models often coexists with tentativeness at making models explicit, and ambiguity as to their scope and status. This is the case even with some of the most scholarly representatives of the genre. Thus, while it would be regrettable if the social history of science were to abandon its valuable empiricist orientation, much would be gained if historians formulated their explanatory tasks in more explicit terms. This would not turn the enterprise into an empty theoretical exercise, but might contribute to *practical* historiographical concerns with the handling and arrangement of concrete materials.

Many possible new approaches and insights for the social history of science can be found in the work of social anthropologists. Work in this discipline, like that in history of science, has been centrally concerned with the study of forms of culture. But this work has been less strongly constrained by any single set of *a priori* epistemological or evaluative assumptions. And such assumptions as it has involved have often differed significantly from their counterparts in the history of science. Cultural anthropologists have not been so frequently or so deeply committed to the forms of culture they have studied as have historians of science. Thus, anthropologists' work is a useful resource in an attempt to broaden perspectives in the history of science.

This chapter is structured so as to point out the possible utility of anthropological perspectives for social historians of science. First it summarizes some contrasted anthropological theories relating interests to knowledge. Then it applies them to a particular body of knowledge which has already been studied by historians of science—

the phrenological model of the brain and mental function. The aim is not to display new "facts" concerning the phrenological movement; indeed, the episode is chosen precisely because reference can readily be made to existing published accounts. Rather, the effort will be to see whether a more-explicit-than-usual analysis of the social epistemology appropriate to understanding this episode may serve to refine available historical techniques for imputing beliefs about nature to social groupings.

ANTHROPOLOGICAL ACCOUNTS OF THE INTERESTS SUSTAINING KNOWLEDGE

(1) INTERESTS IN PREDICTION AND CONTROL

Conceptions of nature in preliterate societies vary widely, and, needless to say, frequently show fundamental divergences from our own "scientific" conceptions. Inevitably, in attempting to account for preliterate cosmologies, anthropologists have related them to analogues in our own thinking. One of the central controversies in social anthropology concerns whether such cosmologies should be taken as analogous to our own scientific theories (and, hence, as responses to the *technical-instrumental* interests which sustain them), or whether they should be treated as responses to *social* interests and preoccupations (routinely considered to be absent from scientific thinking).

Since anthropologists have commonly seen themselves as defenders of, or apologists for, preliterate cultures, it is not surprising that the former position is held only by a minority. If preliterate conceptions of nature are analogues of modern science, then surely they are inadequate science, inferior to and presumably less rational than our own; if, on the other hand, they are responses to social interests, there is the opportunity to display them as rational, adequate responses to such interests (Barnes, 1973). Nonetheless, a tradition does exist of treating preliterate cosmologies as analogues of science, and, if we move back to the more self-confident, ethnocentric anthropologists of the 19th century, we find that this was then the predominant interpretation.

In *The Golden Bough* (which now enjoys a certain notoriety among anthropologists) Sir James Frazer characterized magic, for him a typical manifestation of primitive naturalistic thinking, as a kind of proto-science—inadequate science to be sure, but no less animated than our own science by an interest in explaining the natural world and instrumentally controlling it (Frazer, 1960; cf. Tylor, 1871; Evans-Pritchard, 1965). The primitives were, of course, "mistaken" in their beliefs that magical incantations and the like could move the universe to respond in the desired manner, but:

> When all is said and done, our resemblances to the savage are still far more numerous than our differences . . . after all, what we call truth is only the hypothesis which is found to work best. Therefore in reviewing the opinions and practices of ruder ages and races we shall do well to look with leniency upon their errors as inevitable slips made in the search for truth. [quoted Douglas, 1966:24; see also Frazer, 1960:56-57]

This patronizing approach to primitive cultures is of course anathema to modern anthropologists, and has been "corrected" by those moderns who, like Horton (1971) and Skorupski (1976), continue to stress the analogy between preliterate thought and modern science. But one distinctive feature of Frazer's long outmoded work is nonetheless worthy of note. Frazer gives preliterate beliefs a context of use. However inadequately, he displays the beliefs being brought to bear upon problems of the prediction and control of nature. His famous laws of "sympathy" and "contagion" are held to operate in particular contexts as aids in the manipulation of natural processes. One could be forgiven for applying a later terminology to Frazer and taking his account as an attempt to make out preliterate cosmologies as sustained by interests in prediction and control (cf. Habermas, 1971).

For all that Frazer's influence on anthropology is said to have been "a baneful one" (Douglas, 1966:28), a so-called "neo-Frazerian" movement in the last 15 years has thrown interesting new light on the comparative study of modern and primitive thought.[2] The most influential of these "neo-Frazerians" is Robin Horton, whose "African Traditional Thought and Western Science" (1971) has been surprisingly well-cited by historians of science. Horton takes preliterate cosmologies as genuine attempts to understand and explain nature, in the same way as scientific theories are. And he does not deny the existence of important differences between them and science. But he takes care to avoid devaluing or patronizing preliterate thought, and emphasizes its essential rationality and its close similarity to modern scientific thinking. In this he is helped by an unusually detailed knowledge of modern natural science, which helps him to avoid the common but unfair practice of comparing the reality of preliterate thought with an idealization of modern thought.[3]

As an example, consider Horton's treatment of a phenomenon frequently found in preliterate cultures, that of a detailed isomorphism between their conceptions of the natural order and the social order. It is often said that preliterate cultures use their own society as a model for structuring nature; they are thus guilty of thinking metaphorically and hence irrationally. Horton agrees that the organization of society is often used in this way as a *model* for the organization of nature. But, he insists, this explanation of the unfamiliar in terms of the familiar is a universal and natural mode of thought (Horton, 1971:223-225). It is typical of thinking in the natural sciences where, for example, atomism is an instance of familiar phenomena being used to model unknown or unfamiliar ones. For Horton the social order is a resource, like any other, which men may use to construct cognitive systems. The social order *is* so used in traditional belief-systems; hence, isomorphisms and homologies between the social and the natural are to be expected, and are the products of rational thought.

For an anthropologist, Horton provides us with an unusually abstract, contemplative view of culture. He tends to compare science and preliterate thought as formal abstract systems of beliefs, commenting upon their similarities and differences as sets of ideas and patterns of abstract inference. In neither case does he give us any real insight into the *context of use,* into the practical employment in particular cases of these forms of culture. The beliefs of science and of preliterate cosmologies are treated as abstract pictures of reality performing no function other than that of satisfying abstract intellectual curiosity—a suspect diagnosis in both cases. Unlike Frazer, Horton does not treat culture as sustained by an active interest in prediction and control.

(2) SOCIAL INTERESTS

The "Frazerian" tradition is a minor, albeit valuable, strand in contemporary British anthropology. The dominant "functionalist" tradition, in contrast, insists upon important differences in the interests and functions which sustain preliterate and scientific cultures. The former is predominantly a response to, and sustained as a response to, social concerns.

An important writer in this tradition is Mary Douglas, who has explored several versions of the "functionalist" thesis over a number of years, and who has recently extended her accounts of how cosmologies are the products of social interests to apply to all cultures, preliterate or modern, anthropocentric or scientific (1966, 1970,

1975, 1978). Douglas's later work anticipates several of the themes of this essay (see Barnes and Shapin, 1977).

Her thesis concerning the growth and maintenance of the cosmologies of preliterate cultures is this (Douglas, 1966:ch.5): people, universally, have interests in social management and control. They try to maintain institutions that serve them well; or they actively try to criticize and discredit institutions which work against their interest. Individually and in their various groupings they are always attempting to persuade each other to do what they want. In the course of this activity they invent, invoke and evaluate conceptions of the natural order, which in many cases become institutionalized. Thus knowledge of nature is sustained by social interests, which have a bearing upon the processes via which the knowledge is constructed, judged, and institutionalized. Where witchcraft beliefs, for example, are developed, institutionalized and sustained by a society which employs them to negotiate fluid, unstructured, and endlessly threatening and problematic relationships between peers in the society, the beliefs are to be explained in terms of their *use* in negotiating the relationships.

Thus, with Douglas, it is an understatement to say that she displays the *context of use* of beliefs and cosmologies. The beliefs themselves are only intelligible as they arise out of a context of use, and are credited with no independent abstract existence at all:

> People are living in the middle of their cosmology, down in amongst it; they are energetically manipulating it, evading its implications in their own lives if they can but using it for hitting each other and forcing them to conform to something they have in mind. . . . People are using everything they can . . . to influence one another. [Douglas, 1975:60-61]

Hence Douglas takes a very different view to Horton concerning isomorphisms between conceptions of nature and conceptions of society. Society may indeed be a model used to construct natural realities. But why do people use *that* model? Horton tends to the view that in preliterate societies their own organization is the only model of order they possess, their only real cognitive resource. Douglas, in contrast, emphasizes that that particular model is apposite where social interests are at stake. As part of the cultural resources which people use to deal with practical problems, a morally alive, personalized, responsive nature will "crystallise in the institutions." A natural reality which is "like" man, and "like" the society in which he lives is created "as the appanage of . . . social institutions" (Douglas, 1966: 91). Natural classifications are "like" social classifications because the former are used practically to reinforce, justify, and legitimate the latter. It is a *practical* interest in persuasion and control which underpins Douglas's social epistemology, not a contemplative one. But there is no necessity that "nature" will in all respects, at all times, be "like" the social order. People do not just justify; they may also seek to discredit and undermine social institutions. *Institutionalized* cosmologies will tend to reflect the practical concerns of the dominant institutions. However, discontented social groupings may elaborate cosmologies which are not "like" the dominant institutions in which they live, but which are, rather, signals of an ideal social order or tools crafted to subvert the dominant order. Douglas challenges Horton's contemplative view of preliterate cosmology with the fact that it is not *always* isomorphous with the social order, even in the simplest cultures.

There is, however, an equally significant body of writings which relates beliefs about nature to social interests and concerns, without Douglas's stress on their actual employment and their particular context of use. This is the account of preliterate cosmology as *expressive symbolism* put forward by Firth, Beattie, and many others of the "Oxford school" of anthropology. For Beattie, it will not do to treat the "magical" beliefs of the preliterate cosmos as a kind of "pseudo-science" (Beattie, 1970:245). *Pace* Horton, to do this is to debase the worth of tribal man's intellectual productions. The magical act is so "obviously" a failure as an attempt at control that to attribute "belief" (cf. Needham, 1972) to the primitive is to insult him. What, then, *are* magical rites and beliefs if they are not attempts at explanation and control?

Beattie's answer is straightforward: there is a "crucial" difference between "'ritual' procedures and so-called 'practical' or 'scientific' ones." When we encounter recognizably "practical" beliefs and behaviours, it is appropriate to ask, "what do they do?"; but when we encounter manifestly inefficacious beliefs, such as those involved in ritual and magical acts, it is appropriate only to enquire, "what do they say?" (Beattie, 1966:60). We should, following Firth (1952), avail ourselves of dramaturgical or artistic analogies. Magic is to be understood as we usually understand art or theatre in our own culture, viz., as an *expressive* act:

> when we speak of ritual we are speaking of something which is basically expressive, even dramatic, whereas when we speak of science or scientific activity as such, however "primitive," we are not. [Beattie, 1966:60]

Modes of behaviour and their underlying beliefs are either "characteristically expressive and symbolic" or they are "characteristically nonsymbolic, and . . . based on observation and trial and error." The former we are to call "ritual or art"; the latter "technology, 'common sense', or 'science' " (Beattie, 1966:63). The divide is basic and absolute.[4] We are distinguishing characteristically different "states or attitudes of mind"; "modern science," in this analysis, is "wholly different" from the world of the magician; the two are "quite opposite ways of looking at and coping with the world" (Beattie, 1966:63, 65). This basic difference consists in the different "interests" which inform art and magical belief, on the one hand, and science and technology, on the other.[5] Thus, the personalized, "thou-like" primitive cosmos is a function of the expressive "interest" in its production. Modern science, in this view, has completely shed any such expressive "interest" and, thus, *corresponds* to reality.

How might an "interest" in expression affect the body of knowledge produced under its influence? How might such an "interest" explain links between social structure and cognitive content? Firth's analysis of primitive art is Beattie's paradigm of the workings of an expressive "interest." The primitive artist, says Firth, produces obviously "non-naturalistic" artifacts. This shunning of "realism" is the effect of the symbolic load carried by primitive art. The artist is not primarily interested in depicting or communicating reliable information about the natural world so as to provide a secure basis for practical action. Rather,

> In general, primitive figure sculpture is concerned to bring out certain social attributes of the figure or to express through it certain sentiments which are of importance in the culture of the people. To this end, no exaggeration or distortion is amiss. [Firth, 1952:174]

Primitive sculpture (and hence primitive belief) is "not an abortive naturalism"; to say so would involve a danger of Frazerian recidivism. What the primitive artist "does in many cases is to select and represent what may be termed the social proportions of a subject" (Firth, 1952:175). Similarly, Beattie asks *what* is symbolized in ritual acts. Many aspects of the social structure may be expressed there: differences in social status, the location and distribution of political power and authority, the social order itself, some segment of it, and so on (Beattie, 1966:66; Firth, 1952:177). The effect of an "interest" in social expression serves to shift representations out of *correspondence* with natural reality; it "involves to a large extent what may be called a shifting of the index-pointer of reality" (Firth, 1952:225). Hence, the divide between knowledge produced under its influence and that produced under the influence of a practical, instrumental interest is total (Firth, 1952:215, 225; 1964:236).

It only remains to remove the inverted commas from an expressive "interest." What *kind* of an interest is being posited? Beattie's formulation is quasi-psychoanalytic; rituals, as, for example, those of the Cargo Cults, serve to *console*. The act of expressing the social predicament in itself is alleged to provide some sort of psychic balm; people "in times of stress" have recourse to the "consolations of make-believe" (Beattie, 1966:71; cf. Firth, 1952:181). Presumably, the mere reflection of a social predicament onto another plane is a way of coping with it; or else there may be some sort of psychic release theory underlying Beattie's account.

In the event, Beattie himself prefers not to speak of expressive acts in functional terms (Beattie, 1966:61, 67). But, in the cause of a more active orientation to symbolic expression, we will want to go further to talk of the purposes the symbolism serves. Picking up incidental remarks in Beattie and Firth (see esp. Firth, 1952:239), we can construct a picture of symbolic expression being employed to signal alliances, communicate intentions, locate the boundaries of group membership, and the like. Such things are done not because of aesthetic urges, or for psychic release, but as a part of practical action. When Firth and Beattie informally move to consider the functions and interests shaping and sustaining expressive symbolism, they tend to lead us back into the world of Mary Douglas again.

Firth and Beattie differ interestingly from Horton over what central functions or interests sustain preliterate cosmologies. But in another respect the views of the three show a remarkable resemblance. Horton has theoretical structures passively depicting nature and thereby providing intellectual satisfaction. Firth and Beattie have representations passively symbolizing society or aspects of society, and presumably giving aesthetic satisfaction. Neither viewpoint gives any stress to *context of use,* to what is being done by representations. However, Firth and Beattie, unlike Horton, recognize the importance of this question. Within the functionalist tradition in anthropology, it cannot be set aside. Thus they seek to explain *why* society is symbolized in the preliterate cosmos, where Horton does not see the relevance of the question.

It would be possible to go on analysing anthropological perspectives on social epistemology and human interest in knowledge. But present purposes may be served by the major perspectives characterized above. Our major concern is to connect these anthropological orientations with historiographical practice in relating natural knowledge to social context. This concern will be furthered most effectively if we move to consider a practical problem in the historical understanding of a particular episode, linking, so far as possible, historical treatment with the various anthropological perspectives discussed here.

PHRENOLOGY IN EDINBURGH

It may reasonably be said that to *describe* phrenology as a social phenomenon would be to *interpret* it in terms of the various interests in knowledge just summarized. It is indeed extremely difficult to start with a straightforward account of the phenomenon without already assimilating it to one or another of our anthropological conceptions of the links between the social and the cognitive domains. There is no way around this, so we must signal the problem and proceed anyway.

As phrenology has recently been the object of intensive study by historians with a variety of orientations, there is no need to retail more than the basic features of the beliefs and their setting (for a bibliographic review, see Cooter, 1976). The system properly called phrenology derives directly from the work of two Viennese-trained physicians—Franz Joseph Gall and Johann Gaspar Spurzheim, who did the bulk of their research in France at the end of the 18th century and in the first decades of the 19th (Temkin, 1947). Their doctrines proved fairly fashionable in the *salons* of Napoleonic Paris, but achieved their real success when exported, under the aegis of Spurzheim, to Britain and America (De Giustino, 1975; Davies, 1955). The leading British convert to phrenology was an Edinburgh lawyer, George Combe (1788-1858), and it was largely through his efforts that the Anglo-Saxon institutionalized form of phrenology was developed. By the 1820s and 1830s phrenology had been established in forums for popular culture throughout the British Isles, and scores of societies devoted specifically to the study and application of the doctrine had sprung up. Conflict between supporters of the new beliefs and the adherents of existing knowledge materialized, especially in Edinburgh during the 1810s and 1820s, when a series of disputes between the two camps erupted (Cantor, 1975; Shapin, 1975, 1978). By the end of the 1830s disputes of this kind no longer occurred—some would say because phrenology had been refuted or discredited; others would say because the phrenologists had been successful in achieving their primary aims. It is not, however, contentious to say that phrenology did not become accepted as an academic discipline, no matter how successfully its interests and perspectives were institutionalized at other levels.

The core of phrenological doctrine was a faculty psychology. The phrenologists posited a number (from 27 to 35 in our period) of distinct and innate psychological faculties; mental function as a whole was orchestrated out of the workings of these faculties, e.g., amativeness, inhabitiveness, tune, and so on. Phrenologists further maintained that each distinct faculty was subserved by a topographically distinct part of the brain, i.e., that there was a cerebral organ of tune which was located in a different part of the brain from the organ of amativeness. The brain was the organ of the mind. Finally, they claimed that, all things being equal, the size of the cerebral organs was an index of their power of functioning. For example, if an individual possessed a big cerebral organ of tune, he would have musical ability in a high degree; if he had a small organ, then he might prove tone-deaf. A corollary to this was that, since the contours of the cerebral cortex were followed by the contours of the skull, one could by observation "read-off" innate mental traits from the exterior of the head; hence phrenology's reputation as the "pseudo-science" of bump-reading. Thus far, we have been as purely descriptive as possible. Let us now attempt to understand phrenology as a social phenomenon, to put the anthropological question to phrenology in a defined social setting.

The setting which has been most thoroughly studied is Edinburgh in the first decades of the 19th century. The starting points for these studies are empirical observations, (a) that beliefs about mental function (pro and con phrenology) varied in that setting, and (b) that, in general terms, adherence to (or opposition to) phrenology was situated among social groups with varying backgrounds and social interests (Shapin, 1975:223-231). Thus, there are available materials sufficient for a project in comparative cultural anthropology. The present discussion will, however, focus mainly on phrenology and its "believers," making them the object of sustained curiosity. The project is to interpret the social phenomenon of phrenology in Edinburgh by relating historical materials and anthropological orientations, and to reflect on what the exercise implies for the history of scientific thought.

We know regrettably little about the setting in which phrenology was originally devised, about the individual motivations of Gall and Spurzheim, and about the acceptance of their system on the Continent (Temkin, 1947; Ackerknecht and Vallois, 1956; Lanteri-Laura, 1970; Lesky, 1970). However, this is no problem as we are here concerned only with why a given group of people (the Edinburgh phrenologists) *adhered* to these accounts. Enough is known to say that both the originators of the system and the Edinburgh group were concerned to explain a number of "psychological," "sociological," and "neurological" phenomena. They wanted to account for individual variation in personality and to ground individual character in the interplay between the innate and the environmental. And they sought to explain how the social order is constituted out of individuals with particular psychic endowments. This led them on to the question of what was the natural basis for innate character traits, i.e., what was the *brain* like which was the organ for the expression of these traits? Both Continental founders and Edinburgh groups used their answers to these problems as guides to *action* in society; so to speak, as intellectual tools for jobs of social work.

Various phrenologists, and various institutionalized forms of the system, put different weights upon psychology, society, and neurology. The Edinburgh group stressed psychological diagnosis and sociological theory, but accepted, and, to an extent, furthered the neurological work largely done by Gall and Spurzheim. Let us, however, proceed to outline the explanations of natural phenomena which the Edinburgh phrenologists provided.

They maintained that one could learn about inner psychic states from external signs. Specifically, they maintained, on "empirical" grounds, that one could discern innate mental constitution from the configuration of an individual's head. All Edinburgh phrenologists claimed that this could be done, and cited many occasions when they had correctly diagnosed an individual's character with no other information than that available to the senses from the contours of the skull. Furthermore, they asserted that such diagnosis was an invaluable guide to practical social decisions, such as choosing employees and wives. They claimed that, all things being equal, an individual's innate endowment of organs of various sizes would determine his character. But they also noted that environmental influences might act to modify the power of functioning of one faculty, or, by stimulating the activity of another faculty, to mask its effect in personality. Thus, they did not claim with absolute certainty to diagnose character from the skull, because there were always environmental factors which might deflect innate endowment from unfolding itself along "natural" lines. Likewise, they exempted the old and the diseased from their diagnostic ambit, as natural deterioration or pathological processes might render diagnosis unreliable.

The modifying role of the environment linked the psychological to the sociological explanatory theory. People's interactions—their social relations and their functioning in economic systems—constituted the major environmental influences stimulating some faculties and depressing the activity of others. *All* men, since the creation of the species, had the same set of mental faculties; hence the differences in cultural systems were to be attributed to differing social and economic environments calling some faculties into greater activity than others at any given stage of human development. Thus, in ruder ages, animal propensities dominated; as society progressed, first the moral sentiments, then the intellectual faculties, were called into play and made themselves felt in the value-systems and culture of a society (Combe, 1840:188-191). Social values and sentiments were thus explained in terms of the interaction between individuals' innate psychic endowments and the institutions of a particular society.

(1) INTERESTS IN PREDICTION AND CONTROL

It is possible to give an account of phrenology in the tradition of Frazer and Horton. Materials appropriate to such an account are available in Shapin (1978). It is not just that its practitioners made phrenology out to be a science and offered proofs of its scientific character. Nor is it only that we can point to a context of use for its doctrines in technical diagnosis and political calculation in order to establish its status as more than "make-believe." We can actually point to its role as guiding theory in anatomical research, and point to currently accepted anatomical findings which we possess as the result of such research. And we can show how phrenological doctrine was developed and modified to improve its technical, predictive adequacy in ways which did not obviously improve or add to its merits as a symbolic expression of social organization (Shapin, 1978).

For present purposes the phrenologists' neurological work offers the best evidence and illustrative material. It is possible to demonstrate the importance of interests in prediction and control in the phrenologists' cerebral anatomical work. They asserted, as already noted, that the cranial bones followed the contours of the cerebral cortex. In reply to objections from Edinburgh anatomists that this was not so, local phrenologists actively furthered the Continentals' work. They averred the possibility that one could relate functional areas of the cortex to morphological differences, and, thus, "map" the faculties onto the appearance of the brain surface. In this effort, they represented the convolutions of the cortex with greater care than their contemporaries (even though they did not produce a standard map of the convolutions such as was constructed by the 1860s). They maintained that grey and white matter had distinct functions, and that the main mass of the brain consisted of *fibres*. These fibres were held to be of two types: one type connected the cortical surface of one hemisphere to the corresponding part of another; the other type connected cortical surface to the brain stem. Accordingly, the organs were conceived of as differentiated wedges—fibrous bundles running from convolutions to spinal cord and interconnected by transverse fibres. It was these fibrous wedges which, therefore, provided an anatomical basis for the differentiated organs of the brain and for the mental faculties of the psychological system. In the course of this work on fibres and cerebral regions, the phrenologists also generated and defended against attack some exceedingly detailed work on the pyramidal bodies; the internal structure of the cerebellum; the fibrous connections between ganglia of the cerebellum and brain stem; the appearance of certain ganglia in the brain stem and cerebellum; and the fine structure of the tissues of the convolutions (for full details and pictures, see Shapin, 1978). Although it is not of great significance here, the neo-Frazerian anthropologist (and the Whiggish historian of science) would discover that a great deal of this work is now regarded as part of the body of scientific knowledge, while some of it has been rejected and other parts modified.

Undoubtedly, then, the growth and credibility of phrenological knowledge cannot be understood independently of judgments informed by an interest in prediction and control. But even in the esoteric context of anatomical research on the cerebrum more needs to be said to get anything like a balanced picture. Shapin (1978) argues in some detail that the phrenologists' research on (and "belief" in) their model of the brain was at the same time informed by an interest in displaying the brain as a differentiated mass appropriate to their psychological model. Their commitment to the phrenological "psyche," in other words, informed their attempts to portray a phrenological brain.

That informing interest in their cerebral anatomical work is sometimes readily discerned, and sometimes made apparent only with extreme difficulty, so that much of the phrenologists' anatomy appeared to their contemporaries (as it appears to present historians) as the product of a "pure" interest in explanation and prediction unmixed with any other interest. But when one examines the *context of use* of phrenological knowledge, it becomes clear that the task of defending a theory sustained by social interests led to the production of justifications for it, which would stand when evaluated in terms of an interest in prediction and control. Whatever its scientific merit, the anatomical work was designed to *legitimate* a theory sustained by other factors. This, however, leads us to consideration of phrenology as the product of *social* interests.

(2) SOCIAL INTERESTS

To explain the role of social interests in the development of Edinburgh phrenology we need empirical material on the social basis of that doctrine in the city. Some relevant information is already available (Shapin, 1975). Let us first see how it might be related to explanations following Beattie, treating phrenology as expressive symbolism.

The Edinburgh phrenologists tended to recruit preferentially from bourgeois and petty-bourgeois groupings; this is established both by collective biography of groups like the Phrenological Society and Edinburgh Philosophical Association, and by actors' own perceptions of the social location of phrenological beliefs (Shapin, 1975:226-231).[6] The social position of these groupings within the city around 1800, and the way they related to other classes or strata, has been well documented by contemporary observers, and, for present purposes, their perceptions will suffice (Creech, 1815; Cockburn, 1852, 1874, 1909; [Lockhart], 1819; [Mudie], 1825; Heiton, 1859, 1861; for modern sources, see esp. Saunders, 1950; Smout, 1969). It is worth stressing two general points upon which these observers were agreed. First, they remarked upon the increasing division of labour and specialization of productive function which had been developing in what *had* been characterized as an essentially cohesive and homogeneous social system. It is, of course, well known that academic Scottish moral philosophers were obsessed with the problems of social solidarity and differentiation (Lawrence, 1979; Davie, 1973; Dumont, 1977:ch. 6; Chitnis, 1976:93-115). As Durkheim (1933) would have put it, "organic solidarity" was replacing "mechanical solidarity" in the self-perceptions of these commentators. Furthermore, the building of the Edinburgh New Town, and the resultant decay of the Old Town, struck many as both cause and reflection of growing lack of communication among the classes of Edinburgh society (Youngson, 1966; Heiton, 1859; Creech, 1815). Where before there had been a dominant integrative vision of society, now there were perceptions of individualism, social cleavage and differentiation. While evaluations of these individualizing and differentiating tendencies varied widely, most commentators shared Adam Smith's (1776) view that they were induced by changing material circumstances, particularly in the economic sphere.

Secondly, it was widely believed that traditional social hierarchies and forms of social control were breaking down. Old hegemonies were in decline; new interests and sources of power emerging. Again, many observers attributed these political shifts to economic changes, especially the increasing significance of manufacturing industry and trade in a system which had been dominated by the landed interests and their kin in the professional classes. Henry Cockburn's well-known recollection of a time, in the last decades of the 18th century, when "our merchants were too subservient to be feared" is set against the changed circumstances of the 1820s and 1830s when the commercial classes, in Edinburgh as elsewhere, were more effectively translating wealth into political and cultural influence (Cockburn, 1909: 164-165; Shapin, 1975:223-225). The great efflorescence of Enlightenment culture in Edinburgh had been presided over by the traditional elites of law, land, and established learning. By the early decades of the new century laments were regularly sounded for the decline of the "old thing" (e.g., Cockburn, 1852:i, 156-160; Christie, 1975) and the erosion of the influence of old elites. Commercial interests were becoming more assertive, less timid, more complacent about their future prospects. They were beginning to find their authentic voices and to express their sentiments more volubly; cultural organizations which catered primarily for them were established, as were newspapers and cheap literature. Agitation for the Reform Bill of 1832 crystallized rebellious tendencies which had been increasing for a generation past. A general perception was, thus, that of collapse in the upper part of the social hierarchy—the middle rising up to meet a declining top (Shapin and Barnes, 1976:235-240). Rising bourgeois groups complacently viewed the great transformation; traditional elites searched around for strategies to combat, modify, or adapt to changing circumstances.

Thus, with some necessary oversimplification, we have identified two major aspects of the social world in which the Edinburgh bourgeoisie saw itself situated. Whatever the "true social reality," it is at least clear that the groupings from which phrenology preferentially recruited its adherents perceived it in the way described. Two factors were central to their described social experience: the first was differentiation and individualization; the second an optimistically viewed collapsing hierarchy. Can we then conceive of Edinburgh phrenology as "expressing" such a social experience? There is good evidence that we can and should.

Consider first the phrenological conception of the brain and mental function. One interesting difference between the phrenologists' faculty psychology and that of Scottish Common-Sense philosophers like Reid and Stewart arose from the new system's increase in the *number* of posited primitive faculties each man possessed (Reid, 1812:i, 98ff; iii, 194ff; Stewart, 1854:i, 51ff). Where Dugald Stewart stressed that mental phenomena "are found to be the result of a comparatively small number of simple and uncompounded faculties," George Combe actively criticized the adequacy of a psychology whose "catalogue of intellectual faculties embraces only Perception, Conception, Abstraction, Memory, Judgment, and Imagination" (Stewart, 1854:i, 51; Combe, 1840:176).[7] How could such an undifferentiated model of the mind explain how one man was fit "to be a carpenter, another a sailor, a third a merchant," and so on (Combe, 1840:176)? A far more differentiated model of the mind was needed to explain the division of labour in society, and, more importantly, to account it *natural:*

> The Creator has arranged the spontaneous division of labour among men, by the simplest, yet most effectual means. He has bestowed different combinations of the mental faculties on different individuals, and thereby given them at once the desire and the aptitude for different occupations. [Combe 1840:176]

By varying the degree to which phrenological organs acted in individuals, God "has effectually provided for variety of character and talent, and for the division of labour." Where Common-Sense philosophers of the universities attributed division of labour largely to environment or social arrangements, the phrenologists made it out to be "a direct result [of man's] constitution" (Combe, 1840:177-178). Each individual was distinctly endowed with innate combinations of mental faculties ([Combe] 1817:249):

> Every one differs from another in the size and activity of his organs, and from this source arise the differences of taste. . . . The combination of the different organs, in regard to relative size and activity, determines the particular characters of individuals.

Thus it was that the differentiation and specialization perceived by bourgeois groups was expressed in their model of mental function and human innate endowment.

This psychology and neurology also formed the basis, as we have seen, for a naturalistic view of the social order. The phrenologists' sociology was based on the notion of the *moral individual.* The *Homo hierarchicus* of feudalism and the *Homo sympatheticus* of academic social philosophy was being replaced by what Louis Dumont has called *Homo aequalis* (Dumont, 1965, 1972, 1977). The social order was an epiphenomenon of *individuals'* natural psychic endowments. It was not fundamental and transcendent as the Common-Sense philosophers averred; it was simply an artifact of the interactions of naturally endowed individuals. Moreover, that phrenological faculty which, more than any of the 34 others, "created" society was that which expressed *competition,* viz. the organ of "Conscientiousness," which made individuals sensible of how they stood competitively *vis-à-vis* other individuals:

> The faculty of Conscientiousness, in particular, seems necessarily to imply the existence of the individual in the social state. . . . The sphere in which Conscientiousness is most directly exercised, is that in which the interests and inclinations of equals come into competition. . . . Conscientiousness is not a factitious sentiment, reared up in society, as many moral philosophers and metaphysicians have taught,—but a primitive power, having its specific organ. [Combe, 1840:171]

This "naturalizing" of Adam Smith is then properly to be interpreted as an expression of the social experience lived by the Edinburgh adherents of phrenology, and also as an expression of quite fundamental social changes affecting the whole of industrializing society. Phrenology was "in the business" of creating the psychological basis for our modern conceptions of the individual "in" society. In that sense, we are all now members of the species *Homo phrenologicus* (Dumont, 1972:38-43).

Let us now see whether the phrenological cosmology might also have expressed the Edinburgh bourgeoisie's social perception of collapsing hierarchies. We should be sensitive here to the implicit use of the human body, and particularly of its organization in structural analogies, to express certain types of social experience (Douglas, 1970). Compare the phrenologists' conception of the relationship between mind and body with these typical formulations from the Common-Sense corpus:

> There appears to be a vast interval between body and mind, and whether there be any intermediate nature that connects them together, we know not. . . . There are two great branches of philosophy, one relating to body, the other to mind. [Reid, 1812:i, pp. viii-ix]

> [It is an error] to explain intellectual and moral phenomena by the analogy of the natural world. [Stewart, 1854:i, 54]

The academic intellectual elite of *ancien régime* Edinburgh may then be seen as "expressing" their perception of an intact social hierarchy, the "head" distinct from, and ruling, the "hand." The boundary between spiritual and corporeal was well-maintained (cf. Shapin and Barnes, 1976).

Basic, of course, to the phrenological cosmos was the perception of a collapsed mind-matter hierarchy, no rigid boundary protecting the "spirit" from contamination by the "body." The brain was the organ of the mind, and the mind could legitimately be discovered by observation of its corporeal residence. For practical purposes matter and mind were one; only for contextual strategic purposes were unambiguously materialistic public pronouncements avoided (Gibbon, 1878:i, 79, 92-93, 211-213, 238, 292). Religious feelings were nothing but the products of the operation of cerebral organs (Gibbon, 1878:i, 224). Not only were the valleys to be exalted; the mountains and hills were also to be made low.

The mind-matter hierarchy was not the only hierarchy the phrenologists collapsed. Man was now, more than in any existing philosophy, definitely part of nature; he was, indeed, a natural object and a natural product. What is more, nature was a *good* nature; happiness followed automatically from conforming to natural laws. Combe's *Constitution of Man* (1828) embroidered this theme at length. "Physiology" took the place of piety:

> Under the natural laws everything is arranged harmoniously with the constitution of man; and our only prayers require to be that we may discover what is right, and act up to it, in perfect assurance that God's blessing will inevitably follow, because He has arranged blessings as the consequences of knowing and obeying His laws. [Gibbon, 1878:i, 224]

Combe's miserable experiences at school convinced him that unhappiness is the inevitable result of placing individuals "in circumstances at variance with their nature" (Gibbon, 1878:i, 28). When apparently inequitable social arrangements caused him distress, he turned to his "intense love of nature" and sought succour outside the city (Gibbon, 1878:i, 44). A congenial nature and a highly esteemed "flesh" might then be seen as expressing the complacent optimism of the emerging bourgeoisie (Bloor, 1976:ch. 4; Douglas, 1970, 1978; Willis, 1975).

That the Edinburgh phrenologists did not see social hierarchies as *totally* collapsed is nicely illustrated by an interchange Combe had in 1840 with a more radically inclined political economist. Combe had always insisted, *contra* the Common-Sense school, that *all* primitive powers of the human mind were in principle good (cf. Reid, 1812:iii, 199). Yet he was inclined still to array the types of faculty hierarchically —the intellectual and moral faculties being more highly esteemed than the animal propensities (which man shared with the brutes). Thus Combe was chastized by W.B. Hodgson for his residual preservation of the spirit-body hierarchy:

> To the supremacy of the Intellect and Moral sentiments, or to the supremacy of any of the faculties,. . . I have long objected; the brain is a republic, not an oligarchy, much less a despotism. [Gibbon, 1878:ii, 118]

It would, one suspects, be surprising to find a bourgeois ideologue of post-Reform Bill Britain providing perceptions of a totally egalitarian social order. Certainly, Combe expresses no such sentiments, and, in fact, provides us a vision of a "Lamarckian" social evolutionism by which society is advancing towards the type in which men with the most highly developed intellectual and spiritual faculties will predominate (Combe, 1840:188-190). At this point we have come to the limits of what can be accomplished by an account of phrenology as expressive symbolism. It will now be obvious that the "social experience" projected by the phrenologists was not "social reality" itself, and that other groupings projected different perceptions of society. Insofar as symbols represent "society," the society they represent is always a blend of the experienced and the ideal: what a group's perceptions are, what its aspirations are, and what its legitimations are. We need to ask why bourgeois groups in Edinburgh described society as they did and "symbolically expressed" it in the particular way that they did.

Beattie's position lacks two key elements necessary to respond to this question. He gives no convincing account of the interests underlying expresive symbolism, and he offers little insight into the *context of use* of such symbolic expressions. One way of filling these gaps is to move to the broader frame of interpretation found in Mary Douglas's work. This more extended perspective does not require any further empirical materials. For example, the homologies between society and nature already outlined would remain interesting, relevant, and sufficiently described; we would merely be encouraged to explore the context of their use, the tasks of justification, persuasion, refutation, and so on, which people attempted to carry out with them. In general, it is merely that an *extra question* is put of the groupings which sustained phrenology.

Douglas would want to know in what practical jobs of social persuasion and management these groupings were engaged; what practical problems of legitimation, control, order, management, and seeking of marginal advantage they faced. May the anthropologist account for phrenology in terms of a practical context of social action? In terms of people's real material interests instead of highly implausible "interests" in aesthetics and psychic soothing?

Again, there is an existing account (Shapin, 1975) which was mainly written from this perspective, and reference may be made to it for much background evidence. The emphasis in that account was not on the social "experience" of the phrenologists, so much as it was on the social *jobs of work* they had to do. As ideologues of the emergent bourgeoisie of early 19th-century Edinburgh their prime task, as they saw it, was freeing themselves from the hegemony of traditional elites: how to undermine the authority of academic philosophy, the domination of Presbyterian clergy, the school and the university? How to formulate new institutions appropriate to their new power, and how to construct new political alliances to consolidate their position *vis-à-vis* the rump session of the old order? In Shapin (1975) the phrenologists' cranioscopy and their empiricism were seen as instruments designed and evaluated as legitimations of a programme of social reform. There was little stress in that account on the homologies between social experience and phrenological cosmology discerned in the preceding section. However, it is a simple matter to assimilate these "expressive" homologies to a more active, instrumentalist perspective.

Consider first the division of labour characteristic of the phrenological cosmology. It does not passively "reflect" the experience of differentiation; rather it justifies it and argues that it is *natural.* The phrenologists are signalling their approval of an evolving industrial system to which the old philosophical elite reacted with distaste and hostility (Davie, 1973). Where Adam Ferguson, Stewart, and Reid took it as their task to devise cultural strategies appropriate to *preventing* the divisive consequences of social differentiation, George Combe's enterprise was that of showing social differentiation to be in the natural order of things (Chitnis, 1976:115). Thus, "nature" legitimated a new system of social and economic relations which served the interests of one group more than it did another. Labour itself, which to the Kirk was the punishment for original sin, was accounted natural:

> The first duty imposed of man in relation to society is *industry*. . . . Many of us have been taught, by our religious instructors, that labour is a curse . . . Labour . . . is not only no calamity, but the grand fountain of our enjoyment. [Combe 1840:173-174]

The idea that labour was evil derives, said Combe, from its unequal distribution in society: "Both extremes are improper. . . . When labour shall be properly distributed . . . it will assume its true aspect, and be hailed by all as a rational fountain of enjoyment" (Combe, 1840:175-176). Likewise, their stress on the *moral individual*—the uniqueness of each individual's natural endowment—may be conceived, not as a passive "expression" of an individualistic social experience, but as a criticism of techniques of social control adopted by the hegemonic old order. Specifically, stress upon individualism emerges most clearly in the phrenologists' criticisms of existing educational provisions—with their catechetical, rote pedagogical techniques (e.g., Gibbon, 1878:i, 17; Combe, 1840:161). After demonstrating how individual character is the outcome of unique combinations of phrenological powers of activity, Combe draws the appropriate lesson:

> No individual is a standard of human nature; and . . . those whom we are prone to condemn for differing from us in sentiment may have as good a right to condemn us for differing from them, and to consider their own mode of feeling as equally founded in nature as we consider ours. [Gibbon, 1878:i, 148]

Neither the socializing techniques of the Kirk and schools, nor the "introspective" method of academic moral philosophy, could survive the acceptance of this sort of natural individualism (Mackenzie, 1836:12). Individualism was then a tool for rejecting the socializing tactics of the phrenologists' antagonists amongst the traditional elites—the teachers and the preachers.

The collapse of hierarchy is amenable to a similar reformulation. Nature is approvingly juxtaposed to "spirit" and "culture" because nature was a legitimating cultural resource in the phrenologists' effort to discredit traditional spiritual and cultural *institutions* in society, specifically Kirk and intelligentsia. Materialism and naturalism were strategies—persuasive resources—designed to further real interests.

In 1838 Cobden wrote:

> How I pity you in Scotland—the only country in the world in which a wealthy and intelligent middling-class submits to the domination of a spiritual tyranny. [Gibbon, 1878:i, 314-315]

But Combe and the Edinburgh phrenologists had long been seeking to erode this spiritual hegemony. The elevation of Nature, and the suppressing of the primacy of Spirit, was the tactic they had adopted to achieve this end. The Presbyterian Kirk, said Combe, "believes the physical, moral and intellectual constitution of this world to be greatly disordered." "We," however, contend that "there is a far greater provision made for human virtue and happiness in the functions and capabilities of nature than is generally understood" ([Combe] 1831-1832:198-199, 203, 207):

> A practical as well as a theoretical conflict is permanently proceeding in society, founded on the two great sets of opinions now adverted to. . . . To all practical ends connected with theology, the philosophy of nature might as well not exist. . . . Nature has been neglected in clerical teaching.

Combe's *Constitution of Man* and *Moral Philosophy* were both primarily designed to show that nature was good, that nature was the only sure basis for an ethical system, and that human happiness resulted from obedience to natural laws and not from appeal to supernatural agencies. Human misery followed from the dominance of the culture of the old spiritual elites:

> [Man's suffering] arises from philosophy and religion having been metaphysical, abstract, imaginatory, and too little conversant with the real nature of man and his wants. . . . Break the spell of teaching only abstract morality and religious feeling from the pulpit, and fairly commence a system of teaching anatomy, physiology, and mental philosophy [i.e., phrenology] as the groundworks. [Gibbon, 1878:i, 218, 220]

Moreover, these rhetorical expressions of collapsed hierarchy did not occur in a social vacuum. The passages from Combe's *Moral Philosophy* so liberally quoted above derive from lectures originally given to an organization called the Edinburgh Philosophical Association—a society of shopkeepers, clerks, and members of the superior working classes (Gibbon, 1878:i, 222). Combe, and other Edinburgh phrenologists, were likewise zealous advocates of the scientific (and phrenological) education of the working classes as a whole. As Secretary of the Edinburgh Reform Committee in 1831, it was said of Combe that:

> He was a true Reformer, and sought the solution of the problem which was disturbing the country and its statesmen by the elevation of the lower classes by the means of education. [Gibbon, 1878:i, 251; cf. Shapin and Barnes, 1977:33-45]

Bourgeois Edinburgh phrenologists were thus signalling boundaries of group membership, using cultural resources to seek to discredit an old order, indicating desired alliances in furtherance of their own interests, and elaborating a basis for group action. The homologies between cosmology and society derive from a context of use in persuasion and social management, and not from "dramaturgical" proclivities.

CONCLUSION

Phrenology developed under the impetus of interests in prediction and control analogous to those operative in modern natural science, *and* under the impetus of expedient social interests. The evaluation of phrenological representations was informed by both kinds of interest and one cannot be said to predominate over the other. It is *not* the case that the representations were *assessed* in terms of one interest but *used* in the service of both kinds. The development of phrenological models of the brain, for example, cannot be understood other than in terms of a series of judgments wherein both kinds of interest were inextricably implicated (Shapin, 1978). Phrenology developed as a symbolic expressive/persuasive resource *and* as a body of instrumentally applicable knowledge. Yet there was but *one* form of culture to be identified as Edinburgh phrenology. Phrenology cannot be divided into kinds, or aspects, or even contexts of application so that we can talk of a scientific component, say, and a symbolic or ideological component.

We have a tendency to think of kinds of knowledge or culture corresponding to kinds of interest, and of different interests making incompatible demands upon the way a culture grows. Partly, this is a consequence of our academic arrangements and objectives. In examining culture we tend to have evaluation in mind as well as description, and use the terms "science" and "scientific" as accolades as well as descriptions. Thus the history of science considers culture only insofar as it appears untainted by the involvement of expedient social interests in its evaluation. It has defined its empirical concerns to exclude the role of social interests, and used the term science almost as a definition of culture untouched by such interests (Gillispie, 1959). Consequently, the historian concerned with science is burdened with prejudices which he takes for epistemological divides—an interest in explanation, prediction, and control equals science and technology; an interest in furthering social aims equals ideology. Thus, the unveiling of social interests in what has been taken to be "pure" scientific knowledge is sometimes taken to be an "exposure" of that knowledge and a criticism of its claim to legitimacy. Similarly, anthropologists have been prone to conceive of different interests informing characteristically different bodies of knowledge. The "social function" of beliefs provides a rationale for apparently inefficacious, "nonscientific" beliefs. Functionalism rests upon the same dichotomy between "rational" and "irrational" beliefs, between recognizable "science" and apparent "magic," as the historian intuitively uses. And this again is because anthropologists have been concerned with the evaluation of culture, and not been content simply to describe and explain it.

In practice, the study of how cultural growth is influenced by instrumental-technical interests on the one hand and expedient social interests on the other simply cannot be split into the study of science and the study of "symbol" or "ideology." This is like assuming that electrical and gravitational phenomena can be accounted for in terms of chargeless masses and massless charges. In fact, we have to study charged masses, i.e., phenomena like phrenology.

Forms of culture develop in historical processes over long periods under the influence of different kinds of interest, and cannot be studied as a priori the consequences of one kind only. Detailed study reveals the insufficiency of such an approach. This indeed is why we have stressed the importance of the *context of use* of forms of culture. Investigators who ignore the detailed context of use of culture are liable to miss aspects of the way it develops, and hence are liable to misleading idealizations. Culture is developed and evaluated in particular historical situations. It cannot be studied apart from its use; its use is how we know it. Differentiated kinds of culture, such as we perceive "science" or "ideology" to be, are not concrete embodiments of different kinds of interest: rather they are single sets of resources more or less commonly used and evaluated in particular kinds of context with regard to particular kinds of interest.

It might be said that phrenology is atypical in being so significantly influenced by two different kinds of interest, and that such a situation does not appertain with modern natural science. This *may* in fact be true. Technical-instrumental interests *may* be the only kinds of interest to structure current scientific judgments. But, at present, this is simply a gratuitous assumption, since a possible role for other kinds of interest is scarcely ever looked for. With regard to science, all that we know in advance is that it is a form of culture. Thus it should be studied with an eye for all the possible factors outlined in the current paper. Phrenology offers the general case of how scientific culture should be studied, with an eye to all possible precedents in anthropological practice and elsewhere, and with the burden of demonstration that science is special left with the investigator. To overlook social interests is to misrepresent the history of science; to seek them, successfully or unsuccessfully, is to add to our self-understanding.

NOTES

1. I use the word "epistemology" here in the sense of a theory of how scientific culture *actually is produced,* rather than in the sense associated with some philosophy of science—of how, ideally, it *ought to be* generated.

2. The term "neo-Frazerian" seems mainly to be a pejorative usage among anthropologists in Britain today; hence its surrounding inverted commas, even though no such connotation is intended in the present context. Debates between "neo-Frazerians" and their critics may be located in Wilson (1970: esp. chs. 3, 7, 8, and 12) and Horton and Finnegan (1973). Anthropologists who are reckoned to be amongst the "neo-Frazerians" include (besides Horton) Goody, Jarvie and Skorupski, whose (1976) provides the most sophisticated (albeit philosophical) defence of the "neo-Frazerian" position. A philosopher and historian of science with affinities to the "neo-

Frazerians" is Joseph Agassi (see Jarvie and Agassi, 1967; reprinted Wilson, 1970). For a discussion of these issues and their relevance to conceptions of science, see Elkana (1977).

3. Horton was himself trained as a natural scientist and there are pervasive reminders of this experience in his work (1971). Nonetheless, Barnes (1973) offers significant criticisms of Horton's conception of scientific thought from a Kuhnian perspective.

4. The quotations provided from Beattie and Firth are ample enough to demonstrate that this is, in fact, their basic position. However, in places Beattie sees the desirability of making qualifications, allowing that given acts and beliefs may contain both "scientific" and "expressive" qualities (Beattie, 1966:62-63; 1970:240, 242). But Beattie is neither clear nor consistent; thus in 1966 he claims that "perhaps most" human behaviour "is both expressive and instrumental at the same time" (62), and in 1970 he asserts that only "borderline cases" (like alchemy and astrology) are difficult to classify (242). More important than abstract statements, however, is Beattie's *practice,* and, in this, distinct types of thought and action are characterized by distinct "interests" and modes of thinking.

5. One ought to emphasize that, while Beattie does occasionally employ the term "interest," he is not an "interest theorist." In the main, he opts for formulations which connote the passive and reflective link between knowledge and human experience which he prefers. As I shall make clear below, rendering Beattie's presentation into the language of "interest" exploits his ambiguity but is not intended to misrepresent his own dominant orientation.

6. For some reason, this last point has been conflated with "social determinism"—the position that "all phrenologists were bourgeois" (which is untrue), or that bourgeois status "determines" adherence to phrenology (equally untrue). The point actually being made is that social groupings have distinct interests, which they may pursue in cultural modes, as expressions of their condition and as legitimations for desired states. The recognition that people or groups may have social interests in knowledge helps to account for empirically observed associations between social groupings and "beliefs," but this observation is very far from "social determinism." Nor, indeed, does it have any context-independent implications. If phrenology was a bourgeois ideology in Edinburgh, there is no necessity that it be found to be so in Paris, Vienna, or Boston. Nonetheless, the recent work of Inkster (1977) on phrenology in Sheffield reinforces the Edinburgh findings in a related social context.

7. Phrenology in Edinburgh is properly regarded as an "importation" of a Continental scheme. The Edinburgh phrenologists were themselves intent on depicting their philosophy as different in kind to that of the indigenous academic philosophers of Common-Sense. That two camps were engaged in conflict reinforces (or perhaps even creates) the appearance of bodies of knowledge which were significantly incompatible. Thus, an interesting point emerges if one inspects the two groups' faculty psychology. Reid and Stewart's system posited primitive mental faculties which numbered only slightly fewer than did the phrenologists' original plan; both groups were likewise agreed on the names of many of the faculties. And, when it suited Combe's argumentative purposes, he did not fail to remind the academic philosophers of those similarities (see, for example, Gibbon, 1878:i, 188). Even on the issue of the roles of heredity versus environment and of the material bases of mental function there are often more similarities to be found than the actors themselves chose publicly to recognize (see, for example, Stewart, 1854:iii, 185ff, where he shows remarkable sympathy to "physiognomy"). The phrenologists, therefore, built with materials available in their culture.

The role of "philosophy" in the Edinburgh disputes is not that of a causative agent, but rather that of a tool which actors *elected* to use in furtherance of their interests.

REFERENCES

ACKERKNECHT, E. and H.V. VALLOIS (1956) Franz Joseph Gall, Inventor of Phrenology and His Collection. Madison: University of Wisconsin Medical School.

BARNES, B. (1977) Interests and the Growth of Knowledge. London: Routledge & Kegan Paul.

_____ (1974) Scientific Knowledge and Sociological Theory. London: Routledge & Kegan Paul.

_____ (1973) "The comparison of belief-systems: Anomaly versus falsehood." Pp. 182-198 in R. Horton and R. Finnegan (eds.), Modes of Thought. London: Faber & Faber.

_____ and D. BLOOR (forthcoming) "Relativism in the sociology of knowledge: A scientific justification" in M. Krausz and J. Meiland (eds.), Relativism.

_____ and S. SHAPIN (1977) "Where is the edge of objectivity?" (essay review of M. Douglas, Implicit Meanings). British Journal for the History of Science, 7: 61-66.

BEATTIE, J. (1970) "On understanding ritual." Pp. 240-268 in B.R. Wilson (ed.), Rationality. Oxford: Basil Blackwell.

_____ (1966) "Ritual and social change." Man, n.s. 1: 60-74.

BLOOR, D. (1976) Knowledge and Social Imagery. London: Routledge & Kegan Paul.

CANTOR, G. (1975) "The Edinburgh phrenology debate: 1803-1828." Annals of Science 32: 195-218.

CHITNIS, A. (1976) The Scottish Enlightenment: A Social History. London: Croom Helm.

CHRISTIE, J. (1975) "The rise and fall of Scottish science." Pp. 111-126 in M. Crosland (ed.), The Emergence of Science in Western Europe. London: Macmillan.

COCKBURN, H. (1909) Memorials of His Time. Edinburgh: T. N. Foulis (orig. publ. 1856).

_____ (1874) The Journal of Henry Cockburn (2 vols.). Edinburgh.

_____ (1852) Life of Lord Jeffrey (2 vols.). Edinburgh.

COMBE, G. (1840) Moral Philosophy; or the Duties of Man considered in His Individual, Social and Domestic Capacities. Edinburgh.

_____ (1828) The Constitution of Man. Edinburgh.

[COMBE, G.] (1831-32) "On human capability for improvement." Phrenological Journal, 7: 197-212.

[COMBE, G.] (1817) "Explanation of Dr. Spurzheim's physiognomical system." Scots Magazine, 79: 243-250.

COOTER, R. (1976) "Phrenology: the provocation of progress." History of Science, 14: 211-234.

CREECH, W. (1815) Edinburgh Fugitive Pieces. Edinburgh.

DAVIE, G. E. (1973) The Social Significance of the Scottish Philosophy of Common Sense. Dundee University.

DAVIES, J. (1955) Phrenology, Fad and Science. New Haven, Conn.: Yale University Press.

De GIUSTINO, D. (1975) Conquest of Mind: Phrenology and Victorian Social Thought. London: Croom Helm.

DOUGLAS, M. (1978) Cultural Bias (Occasional Paper no. 34 of the Royal Anthropological Institute of Great Britian and Ireland). London: Royal Anthropological Institute.

——— (1975) Implicit Meanings: Essays in Anthropology. London: Routledge & Kegan Paul.

——— (1970) Natural Symbols: Explorations in Cosmology. London: Barrie & Rockliff, Cresset Press.

——— (1966) Purity and Danger: An Analysis of Concepts of Pollution and Taboo. London: Routledge & Kegan Paul.

DUMONT, L. (1977) From Mandeville to Marx: The Genesis and Triumph of Economic Ideology. Chicago: University of Chicago Press.

——— (1972) Homo Hierarchicus. London: Paladin Pbk.

——— (1965) "The modern conception of the individual." Contributions to Indian Sociology, 8: 13-61.

DURANT, J. (1977) "The Hammerton thesis—a reply." New Scientist, 76 (November 24): 485.

DURKHEIM, E. (1933) The Division of Labour in Society. London: Macmillan.

ELKANA, Y. (1977) "The distinctiveness and universality of science: Reflections on the work of Professor Robin Horton." Minerva, 15: 155-173.

EVANS-PRITCHARD, E. (1965) Theories of Primitive Religion. Oxford: Clarendon Press.

FIRTH, R. (1964) Essays on Social Organization and Values. London: University of London, Athlone Press.

——— (1952) Elements of Social Organization (2nd ed.). London: Watts.

FRAZER, J. (1960) The Golden Bough (abridged ed.). New York: Macmillan.

GIBBON, C. (1878) The Life of George Combe (2 vols.). London.

GILLISPIE, C. (1959) "The *Encyclopédie* and the Jacobin philosophy of science: A study in ideas and consequences." Pp. 225-289 in M. Clagett (ed.), Critical Problems in the History of Science. Madison: University of Wisconsin Press.

HABERMAS, J. (1971) Knowledge and Human Interests. Boston: Beacon.

HAMMERTON, M. (1977) "A fashionable fallacy." New Scientist, 76 (November 3): 274-275.

HEITON, J. (1859) The Castes of Edinburgh. Edinburgh (See also rev. ed. 1861).

HORTON, R. (1971) "African traditional thought and western science." Pp. 208-266 in M. F. D. Young (ed.), Knowledge and Control: New Directions for the Sociology of Education. London: Collier-Macmillan. (orig. publ. 1967; also reprinted in Wilson, 1970: 131-171.)

——— and R. FINNEGAN [eds.] (1973) Modes of Thought: Essays on Thinking in Western and Non-Western Societies. London: Faber & Faber.

INKSTER, I. (1977) "A phase in middle class culture: phrenology in Sheffield, 1824-1850." Trans. Hunter Arch. Soc., 10: 273-279.

JARVIE, I and J. AGASSI (1967) "The problem of the rationality of magic." British Journal of Sociology, 18: 55-74 (reprinted in Wilson, 1970: 172-193).

LANTERI-LAURA, G. (1970) Histoire de la Phrenologie. L'Homme et son Cerveau selon F. J. Gall. Paris.

LAWRENCE, C. (1979) "The nervous system and society in the Scottish Enlightenment." (in this volume).

LESKY, E. (1970) "Structure and function in Gall." Bulletin of the History of Medicine, 44: 297-314.

[LOCKHART, J.] (1819) Peter's Letters to His Kinsfolk (2nd ed.) Edinburgh.

MACKENZIE, G. S. (1836) General Observations on the Principles of Education. Edinburgh.

MACLEOD, R. (1977) "Changing perspectives in the social history of science." Pp. 149-195 in I. Spiegel-Rosing and D. de S. Price (eds.), Science, Technology and Society: A Cross-Disciplinary Perspective. Beverly Hills, Cal.: Sage Publications.

[MUDIE, R.] (1825) The Modern Athens: A Dissection and Demonstration of Men and Things in the Scotch Capital (2nd ed.). London.

NEEDHAM, R. (1972) Belief, Language, and Experience. Oxford: Basil Blackwell.

REID, T. (1812) Essays on the Powers of the Human Mind (3 vols.). Edinburgh.

SAUNDERS, L. (1950) Scottish Democracy, 1815-1840: The Social and Intellectual Backgound. Edinburgh: Oliver & Boyd.

SHAPIN, S. (1978) "The politics of observation: Cerebral anatomy and social interests in the Edinburgh phrenology disputes." in R. Wallis (ed.), On the Margins of Science. Sociological Review Monographs.

______ (1975) "Phrenological knowledge and the social structure of early nineteenth-century Edinburgh." Annals of Science, 32: 219-243.

______ and B. BARNES (1977) "Science, nature and control: Interpreting mechanics' institutes." Social Studies of Science, 7: 31-74.

______ (1976) "Head and hand: Rhetorical resources in British pedagogical writing, 1770-1850." Oxford Review of Education, 2: 231-254.

SKORUPSKI, J. (1976) Symbol and Theory: A Philosophical Study of Theories of Religion in Social Anthropology. Cambridge: Cambridge University Press.

SMITH, A. (1776) The Wealth of Nations. Edinburgh.

SMOUT, T. C. (1969) A History of the Scottish People, 1560-1830. London: Collins.

STEWART, D. (1854) Elements of the Philosophy of the Human Mind, in Sir W. Hamilton (ed.), Collected Works. Edinburgh.

TEMKIN, O. (1947) "Gall and the phrenological movement." Bulletin of the History of Medicine, 21: 275-321.

TYLOR, E. (1871) Primitive Culture (2 vols.). London.

WILLIS, R. (1975) Man and Beast. London: Paladin.

WILSON, B. [ed.] (1970) Rationality. Oxford: Basil Blackwell.

YOUNGSON, A. (1966) The Making of Classical Edinburgh. Edinburgh: Edinburgh University Press.

3

THE POWER OF THE BODY: THE EARLY NINETEENTH CENTURY

Roger Cooter

According to a Roman legend (Ossowski, 1963:91) Menenius Agrippa, when dispatched by the patricians to the camp of the rebellious plebians in 503 B.C., told them a fable about the parts of the body in revolt against the belly. It is said that Agrippa was able thereby to convince the plebians that social classes are mutually dependent on one another, and to have induced them to return to the city.

The legend ideally serves to introduce both the subject matter of this paper and the theoretic problem it treats. Besides illustrating the great antiquity of the analogical exploitation of man's most available metaphor—his body—the legend focuses attention on the particular organismic configuration of the metaphor. We see it here typically deployed as a conservative ruling-class response to social and ideological threats through its dictation of ideas about an immutable kind of social unity. In the 19th century the metaphor was scientized into social theory, eventually evolving into 20th-century "functional-

AUTHOR'S NOTE: This paper has its origins in one delivered to the History of Science and Medicine Seminar at University College London in November 1976. I am grateful to the participants in that seminar for their tolerant comments. The assistance of the following persons who reflected in various useful ways on subsequent drafts of the paper is also gratefully acknowleged: Mike Fellman, Karl Figlio, Bill Luckin, Dorinda Outram, Roy Porter, Bob Young, Chris Lawrence, and the editors of this volume.

ism." This paper intends briefly to explore the genesis and nature of the "proto-functionalist" metaphor as it was popularly deployed in the early 19th century.

The "potency" of the knowledge, or its ability to influence people's thought and action, is the other feature to which the legend of Agrippa lends immediacy. To some scholars popular scientific-medical knowledge is seen in history as simply a cultural resource that does not itself determine activity in any strong sense. Such is the opinion of the editors of this volume who, strongly reacting to traditional idealist sociology in which values are seen to determine consciousness, have gone so far as to express their preferred view (Shapin and Barnes, 1977:60): "that people cannot be controlled through ideas; that it is only through coercion, the manipulation of rights, or the generation of interests that social order can effectively be promoted or broken down." I have no inclination to mount a defence of sociological idealism; I want to suggest, rather, that such knowledge within and inseparable from its socioeconomic context was quite capable of altering people's perception of reality and hence their activity (cf. my view with Barnes, 1977: Ch. 4; and cf. Barnes with Williams, 1977: pt. 2). In particular, I will argue that it facilitated people's entry into and rationalization of the bourgeois social hegemony.

Thus inclined, my argument is in no way assisted by the ostensible effect of Agrippa's fable on the plebians. To attribute such directness to the power of knowledge trivializes the view and makes entirely appropriate a comment like that of Shapin and Barnes (1977:64) on science disseminated in the mechanics' institutes in the early 19th century: that the working-class audience "were evidently as well able as their betters to sniff ideology and reject it." I hope to show that, to physiological knowledge at least, the comment is largely inappropriate; that unlike politico-economic naturalizations at the visible surface of natural knowledge in history, what is truly ideological is deeply mediated and mystified, and that it is at this "scentless" covert level that we must seek the potency of knowledge. But first let us turn to the contents of the physiology and the motives of its promoters, after which we can begin to examine the ways in which it was potent to those who came into contact with it.

"PRACTICAL" KNOWLEDGE

On the face of it the flood of literature written on the body and the laws of health in the early 19th century reflects a growing awareness and concern with ill-health in the rapidly expanding urban industrial world.[1] Written largely by medical men with claims to new resources of scientific-medical knowledge, the literature can be seen as a humanitarian reflex to suffering caused through ignorance. Few authors neglected thus to justify their endeavour, and in most cases there is no reason to doubt their sincerity. Typical was Andrew Combe, M.D., whose highly successful *Principles of Physiology* (1834, and into a 13th edition by 1847) was widely imitated, along with his *Physiology of Digestion* (1836, and into a 9th edition by 1849) and—the work whose popularity was to make him the Dr. Spock of the 19th century—his *Treatise on the Physiological and Moral Management of Infancy* (1840). Samuel Smiles' first work, *Physical Education* (1838), was only one of many such works to acknowledge large debts to Combe. Like most of the polite physiology which was aimed primarily at the swelling middle ranks of society, Combe's works were eminently sensible and readable, offering cogent discussion on the structures and functions of the body and the rational regimen necessary for maintaining health: regular diet, exercise, clean living, moderation in dress and drink, and so on. His works aimed to do good, to help people help themselves and their children to live healthfully and happily in the modern world. Like the man, Combe's writings avoided unction, piety, and pretension.

Yet one has only to stand back from such works or simply take stock of their numbers to have the suspicion aroused that, for all their humanitarianism and innocuousness, their purpose and appeal was larger and deeper than a first glance might have us believe. The sharpest indication of this is to be found in the literature's claim to offer practical scientific knowledge to promote health, when in fact it is merely refashioning common sense under the rubric of newly discovered natural laws. Far from being practical in the sense of, say, John Wesley's *Primitive Physick* (1747), with its 1012 household recipes, the popular physiology of the 1830s and 1840s offers nothing to treat, cure, or ward off disease. From a utilitarian point of view it is useless; for all his own advice, Combe could not stem the tuberculosis that cut short his life. Neither prescriptive nor preventative, the writing of physiology is almost entirely descriptive—descriptive, at the most obvious level—of what was hitherto a mystery for most people, the internal operations of their bodies. This intimate revelatory nature of the science goes far to explain the great demand for it once the popularization had begun. But it is far from being the whole explanation for the knowledge's appeal and popularization. Part of that explanation must be sought in what *was* "practical" in popular physiology: its rhetoric.

Like 17th-century mechanical philosophy, popular physiology culturally expressed a world of scientized common sense in reaction to an image of aristocratic and academic elitist indulgence. Understandable, empirical, and ostensibly practical, it stood for everything that the old order apparently stood against. This was not only implied in the knowledge; it was frequently articulated in ways common to the early 19th-century revivification of dissent. Its authors are not only identifiable as among the upwardly mobile "new men" who filled the literary and philosophical, the phrenological and the philomathic societies that reached their apogee between the mid-1820s and mid-1830s, but are identifiable as among the cultural vanguard of reformists within these societies and in society as a whole. Particularly as epitomized by those in the medical profession, the popularizers can be seen as figuring prominently in this counter-culture out of a heightened sense of ambiguity in their environment. Socially, they lacked any of the real power connected with the production of capital in the new order, and in their own lives most of them were searching for or had only just found economic and social security (Inkster, 1977). In addition, we can speculate that this ambiguity was intensified through the nature of the medical calling: a distribution of dubious physic and the practice of savage surgery in the name of humanity. The searches for certitude in society and certainty in themselves are among the deeper reasons that lie behind the popularizers (and their audiences) seeking satisfaction in physiology, and behind their seeking to popularize it as means of urging on the emergent bourgeois industrial order. Such was the practical rhetorical end which physiology served.

AS RATIONALISATION

Popularized physiology did not seek to rationalize this new order by systematizing bodily symbols or drawing direct analogies. One does not find parts of the body in the literature depicting class boundaries or naturalizing the industrial division of labour. Much may be *imputed* to the repeated importance attached to certain parts of the body in the literature. The heavy emphasis on the stomach, for example, as the "middle" organ whose transformation of raw materials sustained the system, and the claim for the stomach as the only organ to be perfectly operational at birth, suggest naturalization of "middle-classness" quite in keeping with the interests of the advocates of physiology. Similarly, much could be made of the exclusion in most of the texts of any discussion of the reproductive system (which, incidentally, was potentially the most practical point for discussion). But these kinds of emphases or their lack do not reflect symbolization of the social system. They reflect, rather, new interests in the importance of some functional relations over others.

Fundamentally, it is wrong even to search for patterns of symbols in the literature, for it is only in hindsight that we are led to expect them to bear singular meanings for modern industrial society. The popularizers of physiology were not upholding a fixed social reality but were endeavouring to come to terms with the rapid changes in reality due to the emergence of the still inchoate and flexible order. For the 1820s and 1830s especially, we must visualize the popularizers more as "ideologues of change" than mere defenders of the bourgeois structures and relations.

Thus it is not primarily lines of social gradation that are rationalized in the physiology, but change itself—albeit a certain kind of change. This, after all, is why physiology rather than the more certain science of anatomy was the subject of interest. Anatomy had the ability to display the beauty of static physical structures, but physiology could reveal the whole complex structure in active motion (Lovett, 1851:xv). Here, of course, we have a more obvious explanation for the concentration on the digestive organs: nowhere else in human physiology can a regulated dynamic model be so readily illustrated and understood—the familiar upset stomach or constipation speaking volumes for believing in (and attending to) the naturalness of equilibrium and regularity. The laws of digestion literally represented the natural laws of the universe.

What popular physiology offered at a moment in history when the new social relations of urban industrial capitalism were beginning to be manifested in the body politic were compatible images for social organization from which to draw "laws of life." While urbanism and industrial capitalism were fragmenting an older social unity (or impression of unity) and replacing it with the more independent alienated structures of modern society, physiology was presenting holistic images of cohesive parts in dynamic interaction. The body with its differentiated functions was structurally unified, thus revealing an order in nature of cooperation between different individual parts—the emphasis being on the cooperation between specific actions rather than on the parts. Absent in the physiology were those planetary and corpuscular billiard-ball models and metaphors rooted in mechanics, physics, and chemistry which had better suited the rhythms of an earlier age. Instead of the eternal cyclical return of the planetary system or the competitive colliding social atoms of the corpuscular philosophy, early Victorians were confronted with the organically based laws of life—of growth and development, of biology, adaptation, structures, and functions—of organisms living and growing not in chaos but according to the harmony of regulated natural laws. As one author (Anderson, 1826:419) instructed parents on what their children should learn from the contemplation of their constitutions, it was, above all,

> the order and regularity of nature, with the changes that are perpetually taking place in it — the correspondency, the sympathy, the harmony, and the remarkable proportioning of one thing to another, which reigns throughout the universe.

Regularity yet change, order yet progress; this was the service of the organismic metaphor. In a period in which men and women were frightened by the strange social disorders and reverberations that resulted from the same technological and commercial growth that excited them, there was reassurance and promise in a conception of eternal progress combined with natural harmony. The revealed laws of life were all the more comforting if one was casting off the religious yoke of the old order or merely witnessing its erosion. The rationalists who found this revelation most satisfying and were most anxious to promote it were, as noted above, the same who were gaining culturally and socially through the industrial reorganization of society. What physiology held out to them was a naturalization of a meritocratic world view. Specialization being revealed as a fact of nature, it was clear that those who worked in accordance with the (neo-Aristotelian) law of life, that organisms seek to fulfill their essential nature, would be justly rewarded.

The fact that the bulk of mankind was among the masses of labouring poor reflected less that their capacity was fulfilled in manual labour, than that most manual labourers failed properly to bear the stewardship of their natural gifts—they failed to attend to the naturalized Smilesian values. The truth of this was certified on the basis of the celebrated few who, through thrift, initiative, and hard work, managed to climb up from the manual working class to the respectable bourgeoisie. These few also served to sustain the myth that all men actually had equal opportunity to fulfill their natures. As for the idle aristocracy who lived off inherited wealth and land expropriated from the people, they had no place at all in the organismic scheme of things, except perhaps as atrophied parts to be cut out.

The organismic metaphor also validated laissez-faire philosophy, for in natural bodies passages blocked or interfered with meant death or starvation to vital parts. However, by the same token, any number of interventions were also legitimated: what was natural often had to be protected from harmful effects or treated in various ways to restore normal function and the voluntary submission of some parts and functions to others. "Nature red in tooth and claw," and other more fatalistic Darwinian views and rationales for survival in the late Victorian industrial "jungle," were at this time still in the offing. Hence many bourgeois reformers, including Andrew Combe (1832-1834), demanded, for example, that labour's over-exploitation in manufacturing be halted as an infringement of organic laws of life. Behind this was a desire for having the technical imperfections of the system tidied up, combined with the hope that workers and children might thereby gain the leisure time to study the natural laws of organized life for themselves. It was this overall flexibility of the organismic metaphor to both naturalize change and the emerging-as-dominant bourgeois values, structures, and interests that gave this useless knowledge its special attraction to the secular intelligentsia.

AS MEDIATION OF REALITY

In saying that the physiology validated, naturalized, justified, rationalized, or legitimated the values, assumptions, and beliefs that made up the secular bourgeois ideology of industrial capitalist society, I have taken advantage of frequently used and easily grasped concepts for the sake of historical understanding. Given the survival of industrial capitalism and the permeation of its assumptions into all structures of thought, it is not surprising that we lack an adequate vocabulary for expressing the highly subtle and complex ideological process actually involved in the experiencing of this knowledge in its context. So far I have hardly helped matters by leaving the impression that the knowledge descended from heaven, that it was autonomous, and that it was merely tested and found to be a useful rationale for the social system. Inevitably, this attributes to the popularizers of physiology a high degree of consciousness in their activity. It might even suggest that they were in some sense philosophic Gradgrinds rubbing their hands and chuckling at the thought of disseminating a body of natural knowledge that, when pointed to, would ratify the social relations of the industrial order. We forget in this that men like Combe were neither social theorists, drawing detailed analogies between an ideal social system and organic forms of life, nor capitalists grinding the faces of the poor. Reminded of this, we are likely further to forget that "nice" people, especially "disinterested" professionals like Combe, are as capable as anyone else of corrupting, exploiting, and dehumanizing people through the ideologies they hold and disseminate unknowingly.

Thus do we brutalize history and, in this particular instance, circumscribe an understanding of the potency of natural knowledge. When the meaning and deployment of physiological knowledge are interpreted in terms of naturalizing social forms, historical construction is artifically formalized. In practice, men seldom conceive of knowledge simply as a cultural resource for power, nor consciously exploit it as such. To broaden our understanding here we need a subtler appreciation of how knowledge works: first, of the dialectical relations between knowledge and the social system; and, second, of what is meant by mediation.

The organismic metaphor as a "classless" expression of the reality of the class inequalities in industrial society did not descend from heaven. It arose at the end of the 18th century in the work of French physiologists (Manuel, 1956), especially Pinel, Cabanis, Bichat, and Gall. In retrospect their work represents a momentous departure from dominant 18th-century assumptions. By becoming interested in the physiological and psychological differences and complexities of function in living organisms at the expense of previous physiological interests based on (if not actually exploring) qualities of sameness, these late 18th-century physiologists supplied the basis for undermining the reigning Lockean-based paradigm of men's equality. Despite their connections with Enlightenment thinkers of the kind whose philosophical commitment to men's equality made them reject as improbable the story of Agrippa and the plebians (Ossowski, 1963:91), and despite their retention of faith in the malleability and perfectability of man, their work could be seen as revealing inequality as a fact of nature and human nature, and not just (as with Locke) the consequence of nurture. This revelation, in the wake of the egalitarian "outrage" of 1789 and the emergent urban industrialization which was to put individual talents and skills at a premium, allowed the physiologists' knowledge and its orientation to provide an organizing principle for conceptualizing Western society. Henri de Saint-Simon, Charles Fourier, and Auguste Comte were the first modern social theorists explicitly to apply the metaphor. From them, and in an unbroken tradition of biological extrapolations articulating organicist models (most notably by Spencer, 1851, 1860; Durkheim, 1933; Radcliffe-Brown, 1935), the metaphor emerged dominantly and pervasively in the reigning theory of sociology, social anthropology, and industrial psychology: functionalism (Young, 1971).

The important point is that the physiological interest in differences in organisms, which ultimately fed into organismic social theory, arose *simultaneously* with urban industrial society, not ahead of it or subsequent to it. It is a waste of an historian's time to debate precisely which came first; it is more important to realize that we do not have to vote on the issue. Rather we must realize that the origins of the society and its dominant metaphor were in a dialectical relationship, each feeding off the other and increasingly becoming enmeshed. This was how the organismic doctrine came in the 19th century actually to be an expression of social reality, not to determine it (as attempted by the earlier deployers of the organismic doctrine as an analogue), nor be determined by it (Barnes, 1925; cf. Figlio, 1976).

Understanding this dialectical construction of the knowledge in relation to the social structure it metaphorically served facilitates understanding what is meant by physiology's "mediation" of bourgeois ideology. It will be obvious now that by "mediating" the social structure and relations of urban industrial capitalist society, it is not meant that the physiology simply reflected the social form in the biological, as if the physiology were a neutral isomorph of society. Rather, the physiology was interposed between the actual environment and social consciousness, becoming, as it were, an ideological echo or feedback system in the biological domain of values, assumptions, and beliefs in the social. Ideology thus mediated and mystified in Nature makes Nature ideologically latent; Nature can never otherwise be encountered, since it is always fashioned by people's consciousness (Young, 1977: 77ff). In this sense the physiology was, in the dynamic context of its deployment, "live" ideology and not merely a means to ideological rationalization.

The "value-free" factual substance and arrangement of the knowledge therefore, expresses class interests just as much as do the Smilesian incantations in some of the literature which were supposedly derived from the physiological facts. Physiology viewed in this light, as inherently ideological, makes superfluous the distinction that is commonly entered into when referring to popular science—the distinction between "science" and "scientism," or between what is "real" and what is "false" or "ideological." This distinction was not apparent either to the "inventors" of the knowledge or to its popularizers. In the changing context, the popularizers can be seen as simply coming to experience the importance of the ideological echo in the knowledge through the rationalizing satisfaction they increasingly derived from it.

Oblivious themselves, then, to the ideology they conveyed through mystified mediations in the knowledge, the popularizers continued to see themselves and continued to be seen by others as disinterested humanitarians.

THE TAKING OF THE BODY

Ideology, I have been arguing, does not manifest itself at the surface of historical events, but is deeply buried in the experienced organic relations between men, and between men and their world, which are mediated and mystified in the productions of natural knowledge. This view of ideology presses most urgently upon the reception of physiological knowledge by those for whom the ideology was "false." Amplifying the view is the means to understanding the potency of the knowledge. In advance of this, however, two points of information need clarifying.

First, it needs to be understood that while men like Combe were producing this knowledge to, as it were, affirm their cultural worth over the old hegemony of agrarian interests, they were at the same time using every means to have this knowledge put into circulation among "the people," meaning primarily young people and literate artisans. While the former were to be reached through various private educational programmes and massive quantities of well-sponsored literature aimed at parents and teachers, the latter, who interest us here, were reached through innumerable popular journals and institutions of both middle-class and of their own making. Rightly seen by the reformists as the agents in most organized working-class agitations, artisans,[2] if issued with the right knowledge, would, it was felt, more usefully serve as the agents for the passivity of the lower orders. Never was there a time when the lower orders more required scientific knowledge, observed Dr. John Fletcher in his *Discourse on the Importance of the Study of Physiology as a Branch of Popular Education* (1836), adding (what so many similar downward-looking works left unsaid), that the function of physiological knowledge was to teach people to "live without groundless apprehensions" (24); i.e., to accept the nature of things which the knowledge pointed to as unalterable, universal, and eternal.

Again, however, we should not interpret this as meaning that the popularizers of physiology were overtly conspiratorial in exploiting the knowledge's appearance of value neutrality, or even necessarily meaning that they were repressively orientated. The popularizers' desire to pacify and domesticate workers and keep them in their place was rationalized in the belief that their motives were humanitarian and elevating. Disseminators of physiology to working people felt that they were supplying the "unenlightened" with the rational knowledge with which they could combat the superstitious fallacies and myths through which others (notably priests and demagogues) controlled them. The knowledge of oneself that physiology provided was not to reconcile people to their lot in life, but to reveal to them how, through application and "proper" living, they could, if they wanted, rise above their lot to occupy a deservedly higher niche in society.

Following upon this is the second necessary point of information: that the knowledge was well received by artisans, as much among the self-improving Adam Bedes as among the Felix Holts in danger of revolt. I have detailed elsewhere (Cooter, 1978) the great extent to which phrenological knowledge was taken up by working people in this period and their motives for doing so. The extent and the motives for taking up physiological knowledge were additional to these, not different from them. Suffice to say here that productions like that epitomized by Dr. Thomas Hodgkin's *Lectures on the Means of Promoting and Perserving Health, delivered at the Mechanics' Institution, Spitalfields* (1835) were poorly received: the heart of the most staggering contemporary manifestation of capitalism's irrationality was no place to win applause for excessive moral legitimation of bourgeois values and political economy. Works in the style of Combe, however, many of which were not written with the working class specifically in mind, were greeted enthusiastically on the basis both of their informational "neutrality" and the fact that they were frequently assailed by the religious establishment as materialistic. The dread of the prevailing physic and the antagonism felt by many towards the prevailing religious dictates, in conjunction with the welcomed advice on self-fulfillment proffered by physiology, allowed working people to receive much of its literature and lectures as "liberating." Many felt that such information placed their destiny in their own hands: for the first time in their lives they were not excluded from grace, could strive towards it in *this* life, and be accountable only to themselves. It was an exhilarating illusion of freedom.

Rejecting, then, the physiology that was an overt apology for the social relations of industrial capitalist society and that which blatantly endeavoured to instill normative bourgeois conduct—rejecting, that is, obvious scientistic embourgeoisment—working people were all the more willing to accept physiology that appeared neutral, along with that whose rhetoric gave the impression of extending radical materialism. But it was precisely these motives for embracing the knowledge that were the most thorough sources for the obfuscation of the true class realities of working people.

It is not my intention here to deal with the several facets of the rhetorical appeal of physiological knowledge to working people in general and to radical artisans in particular. It only needs to be pointed out that the radicals' faith in science as a levelling resource directed against aristocracy and clergy was an effective source for their cultural exploitation by the radical bourgeois promoters of science. The willingly accepted rhetoric surrounding physiology heightened the sense of a "barbaric" past in which men were kept in ignorance of the beneficial laws of nature at a time when it would have been more pertinent for working people to have focused their attention on their economic situation as a class. Risking oversimplification of a tremendously complex and subtle transformation, one can summarize that artisans who accepted physiology as a further manifestation of reason and rationality with which to hammer the old order, to leaven the social milieu, and to improve social relations (these included Zetetics, Owenite socialists, secularists and many Chartists), unwittingly became the defenders of the irrational social arrangements being advanced through the new economics. Their materialism was expropriated and coopted by positivists who, relying on the same anti-authoritarian assumptions that had once unified what at times had been a fiercely independent artisan-led Jacobin movement against constituted authority (Thompson, 1968; Williams, 1968), now rendered those assumptions protective of the new progressive industrial order. Failing to perceive the ideological power that Reason had assumed (Lichtheim, 1965:169), artisans became its victims, destined (especially where they were instructors of working-class opinion) to promote and safeguard the Reasonable bourgeois world.

It was at a level beneath (though reinforcing) rationalist rhetoric that the organismic metaphor functioned as a mystifying mediation of ideology. At this level it was not the words that mattered but the implied social assumptions that were concealed within and inculcated through the physiological images. Reference has already been made to the literature's popularization of metaphors of function, growth, harmony, order, and so on, and how these suited the outlook of the secular cultural vanguard. To those in the working class who encountered this knowledge simply as neutral descriptive information about the body, the metaphors did not appear as the products of social consciousness. Insofar as they remained assumptions behind the facts, the metaphors did not appear at all and were therefore nearly impossible to frame for debate. There would in any case have been little to contest had it been possible for the organismic metaphor for social organization to have been spelled out as such, for rather than openly challenging working-class self-interests, it merely eliminated the conceptual space for perception in terms of class conflict and struggle. The popularization of the organismic schemata did not usurp working-class struggle in industrial society, for that had scarcely been formulated in the early Victorian period (Foster, 1974). Nor did the metaphor entail the imposition upon working people of a view of reality in which class struggle, conflict, and equality were unnatural. It indeed came to be the case that that view of reality was accepted through the metaphor, but like the rhetorical emphasis on the positive attributes of accepting the knowledge—individual elevation, progress, and improvement—the organismic metaphor functioned in a diversionary rather than an explicitly negative manner. Thus, instead of looking to the conditions they had in common, the recipients of the knowledge were inclined to focus on the interdependence of specialized parts in natural systems and hence to strive to fulfill their unique individual capacities. The validity of human life, they would begin to perceive, derives not from the individual's relation to his fellows or class, but to society as a whole (Manuel, 1956:67). Older social themes current among the common people, of justice, freedom and violent upheaval, would be progressively eroded. And instead of becoming acutely aware of the relevant dichotomies of social life in industrial society—the economic ones between the exploiting and the exploited—they would be guided directly and indirectly by the true light of biology to recognize as more important—more natural—such distinctions as those between adaptive and maladaptive behaviour, normal and deviant functions, normal and pathological states. Attention shifted to these concepts would lead people to evaluate life not in terms of their mutual exploitation and social exclusion, but in terms of their individual adjustment to inherently bourgeois values. Through this internalization, and not by the direct methods hoped for by many of the popularizers of scientific natural knowledge, people came themselves to police their thought and action in accordance with the bourgeois consensual norms being institutionalized. The more they did so, of course, the more they required the knowledge as a rationalization of the alienation in society to which the knowledge led. Once interest in the knowledge was aroused therefore, dependency fed upon need, and need led to consensus.

At this point it is worth remembering that working people were as unprepared for and unused to the emergent socioeconomic arrangements as were others. Hence a cosmology in which class, conflict, and equality were buried had none of the features of obfuscating socio-economic realities that appear to us in hindsight. While alternative models were lacking in the new context, an harmonious order of things stressing the cooperative interdependence of myriad parts within a unified whole seemed to many to be in line with a long-standing vision of a classless utopia. That the "classless" bourgeois realization of the natural order was to be very far from being an egalitarian millenium would not be readily perceived and would increasingly escape recognition the more one entered into the proto-Comtean telos of progressive equilibrium in which class, conflict, and equality remained outside Reason and Nature.

Certainly it does not require a condescending view of ordinary people as passive vessels hoodwinked by clever propagandists to explain the popular acceptance of the proto-functionalist physiology that increasingly reconciled people to an alienating, inegalitarian, and dehumanizing social system. We can see that, being no less caught up in the disorientating forces of change than was the bourgeoisie, working people had little resistance to being socialized by the knowledge that linked them to the emergent social system through their own biology.

Above all, we must understand that in ways parallel to the dialectical construction of the knowledge, the working class experience of it was in relation to the changing external social forms, i.e., in an ongoing process of ideology interacting reciprocally with changing social reality. Physiology helped to obscure for its recipients the vision of irrationality in their environment necessary for desiring to pursue a unified struggle to change it—obscuring this vision at the same time that interest in their individual selves was being heightened. The need for such a struggle would grow less and less evident as the recipients of the knowledge accepted a contradictionless environment and, with that, the impracticability of social and ideological conflict with the social structures and relations of urban industrial society. Alternatives to the progressive meritocratic order, denials of the claim that progress inevitably means differentiation, and refutations of the metaphor whereby parts relate to wholes rather than to other parts, would all have seemed to be outside the bounds of Reason as made manifest through Nature, in spite of the actual mounting exploitative realities. It would seem that to go against Nature—to go against internal design, against universal truth—would only be productive of disharmony, a waste of human potential, an invitation to retrogressive barbaric chaos and confusion. To go with Nature, however, would be to find one's proper niche in the division of labour and develop one's capacities, and so combine personal success with social progress. Thus it was in a complex interplay between organismic mystifications of the realities of class and corruptions of egalitarian ideology, together with what came to seem positive incentives for accepting the alienating natural philosophy based on organicism, that physiology brought outsiders into the consolidating bourgeois hegemony of industrial society.

CONCLUSION

"Consensus" and "consensual" are words that medical physiology introduced into the English language in the 19th century, specifically in reference to reflex nerve action producing involuntary muscle movement (Williams, 1976:67). That it was Spencer, according to the *Oxford English Dictionary*, who in the 1860s first expropriated these words from physiology to apply them in the explication of social functions, appropriately parallels in a literal way the process indicated in this paper: how through physiology actual social and ideological consensus was covertly propagated in industrial society. We have seen that this consensus was not a result of the mere drawing of convincing anatomical analogies like Agrippa's, but, rather, was achieved through confirmation in the social process of meanings, values, and presuppositions conceived within and becoming inherent to the organismic metaphor as popularly deployed. Thus experienced, the knowledge conveyed by the cultural vanguard of the emergent bourgeois ruling class came to be diffused throughout society as constituting reality. Specific class assumptions about the nature of man and his world entered ordinary understanding as classless conceptions. In particular, the assumption that the ideal society was naturally a classless one of eternal harmonious progress and order became a part of social consciousness.

To have accomplished this diffusion of "reality" in the face of the emergence of an industrial order dependent upon the systematic and gross exploitation of one class by another and the maintenance of sharp demarcations between classes was a remarkable achievement. But by far the greatest testimony to the ideological potency of the organismic metaphor is to be found in the unhampered survival of the economic structure and its social relations. This assertion that capitalism has to a large extent been assured by its fusion with the organismic metaphor is based on more than simply a presumption that people's perceptual inability to question the permanence and superiority of the system *derives from* the early 19th-century marriage of biology and bourgeois political economy. The continued pervasiveness of these assumptions is, in fact, largely attributable to (what is only less direct testimony to the ideological potency of the organismic metaphor) the spread across the academic division of labour of the inherited and variously modified organismic structures of thought (Harrison, 1963; Martindale, 1965; Demerath and Peterson, 1967). Given that the functionalist approach to social reality has been adopted principally by those socially positioned to perpetuate the ruling class and reproduce its mental structures, it is not surprising that the idea derived from organicism of total ideological consensus became so extended and extensive. By 1960 class consensus theory in sociology had reached such credible heights that serious attention was being paid to the declaration that the nemesis of ideology itself had been reached (Bell, 1960)—the nemesis, that is, of nonfunctionalist conflict models that would facilitate the perception of reality as conflict-riven. As it happened, the pronouncement of "the end of ideology" was ill-fated: it was no sooner uttered than exposed as more appropriate to the signalling of the end of its own illusive consensual presuppositions.

More recently, however, the signalling has become confused, the light of its initial veracity refracted. In less obvious ways elaborate neo-functionalist weavings still abound in the marketplace of knowledge. How readily we purchase them can be illustrated by the ease with which a functionalist conclusion might be drawn from the contents of this paper. One would only have to suppose that my argument had been that physiological knowledge effected "social control" in the sense of eliminating class conflict in industrial society. This would be like viewing the argument as the secular equivalent of Halévy's thesis on Methodism, regarding organicism as preventing working-class revolution. Besides wrongly assuming that revolutionary proletarian consciousness existed and that there was a working-class struggle, such a thesis functionally presupposes a norm of social stasis and consensus through integration, and envisions revolution the exception that proves the rule. However, the norm in capitalist society is not equilibrium ("dynamic" or otherwise) but continuous irreconcilable class conflict (Stedman Jones, 1977). The thesis here has been that by denaturalizing in people's minds these actual conflicting class relations, organicism obfuscated the true realities of capitalist society. Thus organicism afforded no basis for solidarity among the exploited. This is far different from saying that organicism actually effected social control (stability, conformity, equilibrium) by eliminating class conflict.[3]

The notion of *egemonia* that lies behind the interpretation given in this paper is inherently one of class conflict. It sees the obscuring of class conflict in the consciousness of ordinary people as the result of mystifications of ruling-class ideology through consciousness-determining encounters (especially with knowledge) in everyday life. Through these reciprocally confirming encounters, "false consciousness" ceases to be "false" as one concept of reality—one view of the world and how to regard and treat men within it—is diffused throughout society (Williams, 1973:9; Williams, 1960:587) obscuring an objective view of the irrationality of the relations between men. Such was the cultural function of the popularized and popular organismic metaphor, its overall structure arising out of the real affairs of men and women and being absorbed into the cognitive processes of society. Thus did its mystified mediations of ideology come to influence and restrict the perception and cognition of the ruled.

Hence to reach proper conclusions upon the ideological potency of natural knowledge in the history of industrial society, it is necessary to comprehend thoroughly the ideological formation of the knowledge at the same time that, as historians, we free ourselves from its tenacious grasp.

NOTES

1. It is not the purpose of this paper to document the actual popularization of physiology in the early 19th century. For the related phrenological branch of this growth, see Cooter (1978). There are no separate studies of 19th-century popular physiology as such, though closely connected material on temperance, hydropathy, sanitation, homoeopathy, hereditarianism, birth control, and so on, all of which confirm the depth of the interest in the body in the 19th century, have been dealt with by various historians. My knowledge comes from a wide reading of the popular journals of the period along with the more formal popular physiological works, principally those by (in addition to those cited in the text and to be found in the bibliography) J.L. Beale, Charles Bray, Amariah Brigham, Charles Caldwell, Anthony Carlile, John Conolly, John Curtis, Joseph B. Davis, John Epps, John Forbes, Sylvester Graham, Augustus Granville, William Henderson, and Southwood Smith.

2. Artisans should not be thought of as separate, above, more "ideological," or necessarily distinct and distinguishable from the unskilled and the "industrial masses" in the period under discussion. E.P. Thompson (1968:259) reminds us that

> in 1830, the characteristic industrial worker worked not in a mill or factory but (as an artisan or "mechanic") in a small workshop or in his home, or (as a labourer) in more-or-less casual employment in the streets, on building-sites, on the docks. . . . There were great differences of degree concealed within the term "artisan," from the prosperous master-craftsman, employing labour on his own account and independent of any masters, to the sweated garret labourers.

3. To reiterate in another context what Stedman Jones (1977) has pointed out to historians of "leisure," we may add that it becomes tautological to set about looking for evidence of "social control" through organicism, since industrial capitalism inevitably means the existence of control in this sense.

As for "social control" through organicism in the sense of ongoing behavioural manipulation within bourgeois industrial society, it should be clear from this paper that it is misleading to seek direct evidence of this. Only through a series of complex abstractions is it possible to measure the power of ideology mediated in culture, the phrase itself being an abstraction.

REFERENCES

ANDERSON, C. (1826) The Genius and Design of the Domestic Constitution with its Untransferrable Obligations and Peculiar Advantages. Edinburgh.

BARNES, B. (1977) Interests and the Growth of Knowledge. London: Routledge & Kegan Paul.

BARNES, H.E. (1925) "Representative biological theories of society." The Sociological Review, 17; 120-130, 182-194, 294-300.

BELL, D. (1960) The End of Ideology. New York: Free Press.

COMBE, A. (1840) A Treatise on the Physiological and Moral Management of Infancy. Edinburgh.

______(1836) Physiology of Digestion considered with relation to the Principles of Dietetics. Edinburgh.

______(1834) Principles of Physiology Applied to the Preservation of Health, and to the Improvement of Physical and Mental Education. Edinburgh.

______(1832-1834) "Factories regulation bill." Phrenological Journal, 8:231-238.

COOTER, R. (1978) The Cultural Meaning of Popular Science: Phrenology and the Organization of Consent in Nineteenth Century Britain. Ph.D. thesis, Cambridge University.

DEMERATH, N.J. and PETERSON, R.A. [eds.] (1967) System, Change, and Conflict. A Reader on Contemporary Sociological Theory and the Debate over Functionalism. London: Collier-Macmillan.

DURKHEIM, E. (1933) The Division of Labour in Society. G. Simpson (trans.). London: Macmillan.

FIGLIO, K. (1976) "The metaphor of organization: an historiographical perspective on the bio-medical sciences of the early nineteenth century." History of Science, 14:17-53.

FLETCHER, J. (1836) Discourse on the Importance of the Study of Physiology as a Branch of Popular Education. Edinburgh.

FOSTER, J. (1974) Class Struggle and the Industrial Revolution. London: Weidenfeld & Nicolson.

HARRISON, R. (1963) "Functionalism and its historical significance." Genetic Psychological Monographs, 68:387-423.

HODGKIN, T. (1835) Lectures on the Means of Promoting and Preserving Health, delivered at the Mechanics' Institution, Spitalfields. London.

INKSTER, I. (1977) "Marginal men: aspects of the social role of the medical community in Sheffield 1790-1850." Pp. 128-163 in J. Woodward and D. Richards (eds.) Health Care and Popular Medicine in Nineteenth Century England: Essays in the Social History of Medicine. London: Croom Helm.

LICHTHEIM, G. (1965) "The concept of ideology." History and Theory, 4:165-195.

LOVETT, W. (1851) Elementary Anatomy and Physiology, for Schools and Private Instruction; with lessons on Diet, Intoxicating Drinks, Tobacco, and Disease. London.

MARTINDALE, D. [ed.] (1965) Functionalism in the Social Sciences: The Strength and Limits of Functionalism in Anthropology, Economics, Political Science, and Sociology. Monograph 5, American Academy of Political and Social Science, Philadelphia.

MANUEL, F.E. (1956) "From equality to organicism." Journal of the History of Ideas, 17:54-69.

OSSOWSKI, S. (1963) Class Structure in the Social Consciousness, S. Patterson (trans.). London: Routledge.

RADCLIFFE-BROWN, A.R. (1935) "The concept of function in social science." American Anthropologist, reprinted in his Structure and Function in Primitive Society (1952). London: Cohen & West.

SHAPIN, S. and BARNES B. (1977) "Science, nature and control: interpreting mechanics' institutes." Social Studies of Science, 7:31-74.

SMILES, S, (1838) Physical Education; or the Nurture and Management of Children, founded on the study of their Nature and Constitution. Edinburgh.

SPENCER, H. (1860) "The social organism." Westminster Review, reprinted in his Essays: Scientific, Political, and Speculative (1901), 1:265-307.

______(1851) Social Statics, or the Condition Essential to Human Happiness. London.

STEDMAN JONES, G. (1977) "Class expression versus social control? A critique of recent trends in the social history of 'leisure'." History Workshop: A Journal of Socialist Historians, 4:162-170.

THOMPSON, E.P. (1968) The Making of the English Working Class. Harmondsworth: Penguin.

WESLEY, J. (1747) Primitive Physick: Or, an easy and natural Method of curing Most Diseases. London.

WILLIAMS, G.A. (1968) Artisans and Sans-Culottes, Popular Movements in France and Britain During the French Revolution. London: Norton.

______(1960) "The concept of 'egemonia' in the thought of Antonio Gramsci: some notes on interpretation." Journal of the History of Ideas, 21:586-599.

WILLIAMS, R. (1977) Marxism and Literature. Oxford: Oxford University Press.

______(1976) Keywords. A Vocabulary of Culture and Society. London: Fontana/ Croom Helm.

______(1973) "Base and superstructure in marxist cultural theory." New Left Review, 82:3-16.

YOUNG, R.M. (1977) "Science *is* social relations." Radical Science Journal, 5:65-129.

______(1971) "Functionalism." (unpublished typescript)

Part Two THE RISE OF NATURALISM

Scientific naturalism, both as an explicitly articulated ideology and as a diffuse style of thought, achieved prominence in the early 19th century and gained in importance throughout the Victorian period. Papers in section one by Shapin and Cooter have already examined naturalistic representations of human brain and body in early 19th-century Britain. But it was not until the 1860s and 1870s that scientific naturalism attained its fully developed form. Naturalism asserted the universal scope of scientific method and procedure, the adequacy of science as a universal deterministic cosmology beyond which no further knowledge or way of knowing exists, and the universality of natural law. It denied teleology, metaphysics, and miracle, and thus effectively excluded and rejected clerical authority and the credibility of religious knowledge.

Recent studies (by F.M. Turner, R.M. Young, and others) strongly support the conclusion that naturalism was the ideology of rapidly ascending new professional groups in an industrializing society. Expert laymen, legitimating their authority in terms of the secular ideology of naturalism, competed with and opposed the traditional sacred authority of clerics. And in this they were ultimately sustained by the newly powerful "bourgeois" groupings whose commercial and industrial activities set them in opposition to the landed interest and its traditional minions.

Roy Porter leads us up to this period with a discussion of cosmogonical writings through the 18th century. In cosmogony, religious and secular modes of thought combined; scriptural historical accounts of earth history were partially naturalized. Religious dogma was adorned with the increasingly naturalistic garb considered necessary to legitimate knowledge at the time. But cosmogony was an unstable form which was never able to command any general acceptance among intellectual elites. It was a transitional form as knowledge moved from a sacred to a secular base, and as society moved from sacerdotal to lay professional sources of expertise. As Porter says, cosmogony's credibility was restricted, since "The ideological guardians were split between clerical and lay elites." And, as the "lay elites" triumphed, so cosmogony's time was over. In the early 19th century, "Guardianship of the traditional, fully articulated theory of the Earth became the preserve of the 'scriptural geologists', marginal figures who abused and were abused by the new leaders of the geological community."

And so we come to the full-blown scientific naturalism presaged by phrenology and brought to fruition by Darwin, Huxley, Tyndall, Clifford, and others. At least, this is where we arrive if we accept that the writings of such a wide range of individuals can be properly characterized as a *movement* within a culture—as bringing them under the rubric of naturalism implies. Such a characterization does raise problems of historical method, and clashes with the particularistic, individualistic assumptions sometimes made by historians. In their discussion of Darwin and social Darwinism, Shapin and Barnes address some of these problems and argue that the historical study of a topic such as "naturalism" is no more problematic than the study of a specific individual. In particular, they stress the difficulty of severing the individual from his culture, or undesirable features of that culture; and they criticize the common historical procedure which seeks to achieve such a demarcation by imputing appropriate inner states of belief or motivation to the individual in question. Darwin himself, it is argued, simply cannot be set apart from the thought of his own culture, to be contrasted as a "scientist" with "ideologists" whose thinking was socially embedded. Nor is it in any way problematic or undesirable to regard Darwin's writings as an aspect of naturalism, and to accept that this pattern of cultural change was evident in the context of worthwhile scientific work as well as elsewhere.

Both Joan Richards' and Brian Wynne's papers reinforce these points about naturalism as a broadly based movement of cultural change with repercussions in a whole range of specialized spheres. Richards describes the strongly naturalistic style of the non-Euclidean mathematics of William Clifford and the equally strongly anti-naturalistic response it evoked among Oxford and Cambridge mathematicians. In this paper mathematical judgments and strategies are shown as intimately connected with their implications for the credibility of naturalistic doctrines, the general social implications of which were clearly apparent to the mathematicians involved. Work such as this surely establishes that judgments in the esoteric professional context of mathematics can be conditioned by features of the wider social setting, and yet it necessitates no particular revision of current judgments upon the merits of projective and differential geometries.

Brian Wynne's study of the response to naturalism implicit in late-Victorian Cambridge physics has many parallels with Richards' treatment of the Oxbridge mathematicians. But the particular interest of his paper lies in its detailed attention to the general social context in which Cambridge physics was situated. He is able to detail the physicists' connections to their general intellectual and social context, and thus to reveal their physical concepts functioning in two kinds of situation. Wynne argues that the meanings of these concepts do not arise solely from their use in one of the two contexts, and that there is no empirical basis for giving the one priority over the other. Hence, strictly, it is unjustified here to talk of "scientific" concepts in use outside science, or even of "unscientific" concepts being used within science; both formulations imply an assymetry unsupported by historical evidence. All that this evidence allows is that the concepts were employed in distinct contexts, and that their significance developed and grew according to their use in both.

4

CREATION AND CREDENCE: THE CAREER OF THEORIES OF THE EARTH IN BRITAIN, 1660-1820

Roy Porter

> Ephraim Jenkinson:. . . the cosmogony of the world has puzzled the philosophers of every age.
> Vicar of Wakefield: I ask pardon . . . but I think I have heard all this before.
>
> —Goldsmith (1766:79-80).

This chapter investigates the rise and fall of the genre of scientific cosmogonies in Britain from the mid-17th to the early 19th century. First I aim to establish what such theories of the Earth were, by locating their roots in Christian Scriptural accounts of Creation, and showing where they diverged in terms both of cognitive content and of the interests they were embodying. The next section then analyses the ideological functions of the leading ideas of such theories, as a prelude to the main discussion—that is, why this scientific genre had such a chequered career. For theorists of the Earth were constantly feuding amongst themselves, and the discipline itself was to be ignominiously superseded by "geology" at the turn of the 19th century.

AUTHOR'S NOTE: For criticisms of earlier drafts of this paper I should like to thank Bill Bynum, Roger Cooter, Barry Barnes, Margaret Jacob, Jack Morrell, Gill Morris, Michael Neve, Clarissa Orr, Steven Shapin, and Bob Young.

Why then was the theory of the Earth a failure as regulator of thought and action? The question cannot be ducked. To pose it is not Whiggishly to dismiss such theories as stupid, but to point out that they did not successfully establish and sustain themselves. Yet it would be facile to expect a decisive answer, there being no objective measure of "effectiveness." Cosmogonies failed to fulfil their own expressed aspirations of providing a reading of Man in natural Creation sufficiently consensual to unite the intelligentsia and to be disseminated with credit to a wider audience. Was it because the sociopolitical interests which had to be naturalized were too tortuous and discordant? Or was the fabric of the theory itself elusive and refractory? Or, alternatively, did dramatically *changing* political circumstances radically transform the demands placed upon a cosmogony?

THE BOOK OF SCRIPTURE AND THE BOOK OF NATURE

The theory of the Earth, as a distinct science, flourished from the mid-17th to the end of the 18th century (Roger, 1973-1974; Collier, 1934).[1] It was an answer to the question: what is the meaning of the Earth?—an answer couched in terms of origins and history. The substance of such theories was crisply summarized by Thomas Burnet:

> A *Moral* or *Philosophick History* of the World well writ, would certainly be a very useful work, to observe and relate how the Scenes of Humane life have chang'd in several Ages, the Modes and Forms of living, in what simplicity Men begun at first and by what degrees they came out of that way, by luxury, ambition, improvement, or changes in Nature; . . . This would be a view of things more instructive, and more satisfactory, than to know what Kings Reign'd in such an Age, and what Battles were fought; which common History teacheth, and teacheth little more. Such affairs are but the little underplots in the Tragicomedy of the World; the main design is of another nature, and of far greater extent and consequence.
>
> As the Animate World depends upon the Inanimate, so the Civil World depends upon them both and takes its measures from them; Nature is the foundation still, and the affairs of Mankind are a superstructure that will be always proportion'd to it. [1684: 246]

Such theories of the Earth were the legatees of the Scripture and metaphysics grounded cosmogonies which had traditionally expounded the Christian truths of Man's history and destiny within Creation

(Allen, 1949; Williams, 1948). In the early modern period, the epistemological credit of these latter had been threatened (Jones, 1936; Baker, 1952; Cragg, 1950; Hazard, 1964). Schisms within the Christian churches, between Catholic and Protestant, and then within Protestantism between Lutheran and Calvinist, Anglican and Puritan, had sundered the unquestioned authority of the Church and Holy Writ. Meantime, scholastic metaphysics were being attacked by the New Science, empiricism, systematic doubt, observation, and experiment. Needing new foundations to support the teachings of traditional Scriptural cosmogony, theorists built on the rock of natural truth (Tuveson, 1949; Willey, 1934, 1940).[2] Truth was to be grounded in Nature, as well as Scripture and the authority of time. Thus the naturalistic theory of the Earth as propounded by Burnet, Hooke, Woodward, Whiston, Keill, and, later, Hutchinson, Catcott, Whitehurst, Hutton, de Luc, and others succeeded the largely Scriptural exegetic cosmic histories of previous generations. Raleigh's *Historie of the World* (1614), Stillingfleet's *Origines Sacrae* (1662), and Milton's *Paradise Lost* (1667) were stiffened with mathematics and mechanics. The governing myths of man *in saeculo* were recast in the garb of the New Science.

These new theories of the Earth chimed in aim and function with earlier evidential theology. Just as orthodox Christian cosmogony had always fused via, vita & veritas—norms and commands for action in the family, the polity, the world predicated on objective truth—so the new theorists likewise believed that to understand the universe was to prescribe behaviour, for Nature was the Divine Legislator's self-policing legal system. In Newton's words, "I received also much light in this search [of Scripture] by the analogy between the world natural and the world politic" (quoted in Jacob, 1976: 14). Scientific cosmogony would also confirm Christianity. The empirical history of Nature—the Book of God's Works—vindicated Moses's veracity, and thereby Moses's Laws. As William Whiston wrote, "nothing will so much tend to the vindication and honour of reveal'd Religion as free inquiries into . . . true and demonstrable principles of Philosophy [and] the History of Nature" (1696: 63).

That is to say, theories of the Earth, unlike many forms of scientific discourse, were a quite *overtly* value-laden mode of natural knowledge, giving meaning to life and directing action. This was because cosmogonies focused on Man—his nature, history, destiny, rights, and duties—as their express subject. As man was *Imago Dei,* so *mundus* was *Imago hominis.* Christian and Classical cosmology had always conceived of Nature as constructed for a human habitat in God's providential plans (Glacken, 1967; Passmore, 1974). Except possibly George Hoggart Toulmin (Porter, 1978a, 1978b) blind environmental determinism was to find no British advocates. Sacred history unfolded as a volume of anthropocentric parables showing God through His servant Nature distributing manna and thunderbolts to Man. Hence, in addition to whatever *masked* and *oblique* projections of human interests theories of the Earth may have carried, their surface programme was that the Globe Theatre played out the tragicomedy of human understanding, choice and action, sin and redemption.

The theories under discussion thus broke new ground, in setting out a cosmogony validated by objective Nature as rendered by the methods of science and Enlightenment epistemology. Yet their commitments were familiar (cf. Becker, 1967). The form was radical, the content venerable. Like previous Scriptural scholarship, scientific theories of the Earth proved Divine omnipotence. Neither Nature nor matter could have existed from all eternity, for they were God's creation, who continued to be their controller, for "God is a God of order" (Catcott, 1761: 4). God had contrived the world rational, orderly, and according to intelligent design. God's hand (wrote Woodward, in a phrase quoted by Buckland over a century later) is "a steady hand, producing good out of evil, the most consummate order and beauty out of confusion and deformity" (Buckland, 1820: 15). Nature was made for Man. Furthermore, science traced the footsteps of Providence: miracles, the sacred harmony of macro-and-microcosm, Intellectual intervention. It proved the world did not act inexorably by deterministic law or the random collision of atoms. Like the rest of Nature, Man was made, rather than self-creating, but he was the "last and most admirable work" of Creative Power (Raleigh, 1614: 22).[3]

By means of a narrative of primordial origins Christian cosmogony had defined the present Universe and directed social forms and choices: "as it was in the beginning, is now, and ever shall be." The divine order of Nature of course confirmed the here-and-now system of public and private life. But, more specifically, theories of the Earth provided natural signposts for Heaven, confirming the doctrine of an independent, timeless, spiritual world, a personal afterlife, considered as reward for present godliness (Justice) or as compensation for tribulation (Faith). For Christians, the *civitas terrena* was no more than a shadow of the *civitas Dei*, but theories of the Earth could offer a scientific glimpse of eternity.

Theories of the Earth unpacked the reason of existence by telling its sacred history (cf. Lowenthal, 1975). Man was now the "great amphibian" on account of the sin of Adam: an Adam temporally and genealogically adjacent. Man's physical environment and his own bodily condition—subject to toil, disease, age, and death—were constant reminders of God's expulsion of the first pair from a physical paradise. Man lay in political subjection—so ran favoured Stuart political theories, such as Filmer's *Patriarcha*—because of the power of paternal monarchy vested by God in Adam and his seed forever (Schochet, 1975; Laslett, 1949).

But the human condition was equally guided by its future physical and spiritual destiny. Seventeenth-century systems of the world were through and through eschatological, and often specifically chiliastic (Lamont, 1969; Capp, 1972; Hill, 1971; Toon, 1970; Jacob, 1976). Natural philosophy might predict the time and place of the Millennium, or Consummation, whether purification would be by fire or water, and how man could best prepare for it (Burnet, 1684; Whiston, 1696; Jacob and Lockwood, 1972).

Naturalistic theories of the Earth thus continued to grind out a traditional ideological *basso continuo* enjoining political, social, and familial obedience, and inherited property distribution. But they also had a new melodic line with new words for a changed situation. They can best be approached by being juxtaposed against recent historiography analysing the ideological face of Newtonianism. Newtonianism has been seen as the programme of the Latitudinarian connexion within the Church of England in an emergent market, capitalist society (Jacob, 1976); as the shibboleth of the victors of 1688; and more generally as the battle-cry of the warriors of the European Enlightenment (Guerlac, 1965; Buchdahl, 1961; Kiernan, 1973; Gay, 1967-70; Jacob, 1977). Newtonian natural theology encapsulated the archetypal aspirations of the Georgian Establishment, conjuring up a stable, law-governed universe which also accommodated miracles and special providences; a world of individualism, but one where self-love and social were the same; a system of Nature, though a Nature which required a Creator and Sustainer. Such a view mirrored the myth of 1688—the Providential Revolution (fanned by the Protestant Wind) which restored order; the Whig philosophy of the relations between King and Parliament, prerogative and common law, and—in the workings of the law itself—the blend of the majesty of justice with the grace of mercy (Hay, 1975).

Newtonianism carried great prestige throughout Enlightenment Britain. Its field attracted all but die-hard traditionalists such as the Hutchinsonians, and later Romantics such as Blake (Kuhn, 1961; Neve and Porter, 1977; Ault, 1975). Not surprisingly, many theorists of the Earth avidly clutched at Newton's skirts (cf. Woodward, 1965; Whiston, 1696; Harris, 1697; Keill, 1698; Whitehurst, 1778).

My argument is that the fate of naturalized cosmogonies differs in two important respects from that of Newtonianism per se. In the first place, Newtonianism became a highly protean and pliant body of analogy—conceptions of universal law and harmony, of attraction and repulsion were adaptable metaphors. Cosmogonies on the other hand fixed upon a unique sequence of events and had to resolve a more highly charged complex of dilemmas. Second, as distinct from the hegemonic kudos of Newtonianism, cosmogonies—even Newtonian ones—suffered a rough passage and early demise. By contrast with the Newtonians, no common loyalties united theorists of the Earth.[4] "Whether Dr. Burnet's roasted egg, Dr. Woodward's hasty pudding, or Mr. Whiston's snuff of a Comet will carry the day," wrote William Nicolson in 1697 with sceptical foresight, "I cannot foresee" (Nichols, 1809: i, 104). From the 1690s theorists ran into a polemical hail—much fiercer than ever faced by Newtonian mechanics or natural theology in general—from empirical natural historians, grub street satirists and certain theological quarters (Porter, 1977: ch. iii; Tuveson, 1950; Redwood, 1976; Nokes, 1975). Benjamin Martin was spokesman for all in deeming that all "World-Mongers Systems and Theories, dissolve into a philosophical Nothing, which want actual and repeated Experiments to support them" (1735: 21; cf. Goldsmith, 1820: 13). Opposition culminated in a massive onslaught around the turn of the 19th century from naturalists, theologians, and literary figures alike (Porter, 1977: introduction, ch. viii; Garfinkle, 1955). The genre of the theory of the Earth lost credence amongst the intellectual establishment, being salvaged only by sectarian religious pundits like Granville Penn, Andrew Ure, and George Bugg (Millhauser, 1954). Meanwhile, naturalists picked up the pieces and reworked them into a new configuration of knowledge, the science of geology, which promised greater credit as both science and ideology (Porter, 1977: chs. vi-viii).

THE MEANING OF MAN'S PLACE IN NATURE'S HISTORY

The bedrock of theories of the Earth lay in giving objective reality to the Christian cosmic drama of human sin redeemed through Christ: the world dealt with imaginatively in *Pilgrim's Progress* and *Paradise Lost* (Hill, 1977). Take for instance the cosmogony of Thomas Burnet—essentially pre-Glorious Revolution and pre-Newtonian

(Tuveson, 1949; Kubrin, 1968; Jacob and Lockwood, 1972). Though modern (i.e., Cartesian) in its science, Burnet's *Sacred Theory* (1681) still charted the irreversible divine itinerary from Creation to Parousia. Man inhabited a ruin: a deformed, sick, decaying, senescent world, the damage wreaked by the Deluge, as the Almighty's punishment for original sin. Redemption would begin through purification of the Earth by fire and restoration of its pristine state at the Millennium. Burnet's theory evokes Renaissance nostalgia for the pastoral Golden Age (cf. Hooke, 1705). It echoes Calvinist views of vice and divine retribution and oozes the Stuart mood of mutability, loss, weakness: a tragic resignation. Burnet's pious purpose is to indict corrupt lapsarian man for such times of woe and tribulation. Adam's sins[5] have been visited upon his progeny in the form of disease, mortality, and the curse of labour, through the divinely induced decay of Nature. Man is presently in thrall to his past. The time is out of joint. But it is the Deity who will set things to rights. Margaret Jacob has evoked how aptly this posture of anxious hand-wringing, this insecure providential Millennialism, expressed the precarious plight of Latitudinarian Churchmen, with their bewailing human weakness in a Restoration polity in which their enemies prospered more than they (1976: 107).

Obviously the evocation of nostalgia, decay, and mortality continued to pervade much 18th-century cosmic thought, meshing with the religious aesthetics of the sublime, graveyard poetry, love of ruins, and the gothick (Nicolson, 1959). Yet the meaning of the Earth underwent remarkable change. The Revolutionary Settlement of 1688-1689 ushered in a claque of ideologues whose self-appointed mission was to celebrate the new age as one of freedom, order, prosperity, and progress in the political, religious, social, and economic domains. Intellectuals such as Woodward, Harris, and Keill, sympathetic to this sociopolitical view, constructed theories of the Earth which vindicated the new human order by rendering it natural. They reportrayed man's terrestrial status as providentially favoured. They explicated the natural order as stable, healthy, and enduring. Thus Woodward wrote that the "grand design of Providence" was the "Conservation of the Globe" (1695: 238) in a "just aequilibrium" (cf. Davies, 1966: 133). Whitehurst was later to praise the permanent physical order of the Earth, deeming the laws of Nature "immutable," "gradual," "regular," "uniform," and "progressive" (1778: 8, 14). George Hoggart Toulmin and James Hutton agreed (Porter, 1977: ch. viii, 1978a, 1978b). The formulation of geological Uniformitarianism in the 18th century crowned the present system of things with a philosophical halo.

The Earth had undergone revolutions in the past—necessary and salutary ones. But its revolutionary career was past. All was now equilibrium, tranquillity, order. The body terrestrial was balanced and stable (Woodward, 1695: 228, 238; Tuan, 1968). The rainbow signalled the new dispensation of the growth of terrestrial stability (Woodward, 1695: 101, 136; cf. Plumb, 1967). Furthermore, the last revolution of the globe had been constructive not retributive. It had been a "Reformation," introducing a new "constitution" "into the Government of the Natural World" (Woodward, 1695: 61). Through the Deluge, wrote Woodward, God had transformed Men from "the most abject and stupid Ferity, to his Senses, to sober Reason; from the most deplorable Misery and Slavery, to a Capacity of being Happy" (1695: 94; cf. 60ff). The new world—no decayed and flawed eyesore—had been adapted by the Divine Governor "at the Helm" to fallen Man's needs. Prediluvial abundance had stoked sin (Woodward, 1695: 85). But God had rendered the post diluvial Earth niggardly, so that Man had to labour by the sweat of his brow, thus inducing frugality, sobriety, discipline, and industry (Woodward, 1695: 61ff). Eighteenth-century theories of the Earth drowned down the myth of the Golden Age in choruses of the Protestant ethic and the spirit of capitalism. The Earth was Man's to enjoy—as William Phillips could still boast at the beginning of the 19th century, "*everything* [was] *intended for the advantage of Man*," the "Lord of the Creation" (1815: 193, 191; cf. Passmore, 1974). The globe was a habitable world, perfectly adapted to Man's condition (Hutton, 1795: i, 17-18).

Thus the new theories of the Earth married Locke to cosmogony; possessive individualism inherited the Earth (cf. Macpherson, 1962). The rights of Man to be Lord over the Earth and all its creatures; and to multiply, populate, and possess it as private property were reendorsed (cf. Woodward, 1695: 250f). Goldsmith depicted the Earth as a mansion for Man, God's tenant, to enjoy (1820: 1). Hutton saw it as a farm (1795; cf. Bailey, 1962).

In other words, Enlightenment cosmogonies celebrated the cosmic (and hence by implication the political, moral, and social) status quo. God had chartered terrestrial felicity by his glorious revolutionary settlement. Elizabethans had anticipated mutability, disorder, and prodigies in the natural world as harbingers of human disruption. Stuart chiliasts, including Burnet, had expected cosmic dissolution portending the Last Things and the divinely instigated overthrow of Satanic tyranny (cf. Jacob, 1976: ch. 3). But from the 1690s the natural world had become stabilized. On the one hand, cosmogonical ideologies construed this optimistically as an opportunity. Despotism, original sin, and their correlate the Ruined World, had been dissolved away. The world garden could be enjoyed. But they also spelt out its sterner injunctions to fall in line with reality. The stability of the terraqueous globe pointedly precluded revolutionary and millennial events in the future, and thus put radicals' noses out of joint. Both pious Christians like Ray and Woodward and later Deists like Hutton saw no prospect of an end in the terrestrial machine, just as Adam Smith was to think the economic system indefinitely self-regulating, and Hume and Burke were to champion a naturally stable political order.

The actual constitution of the Earth was here to stay: self-sustaining, self-repairing, self-improving. Recourse to an imminent, millenniary transformation of the terraqueous globe was now foreclosed against radicals, sectaries, and malcontents. The Hanoverian theory of the Earth blinkered men to see only the historical past and the present. The link between past and present in the natural world was a transition from rudeness to refinement, just as contemporary social theory constructed the evolution of civilized man (Whitehurst, 1778: 8, 75; cf. Crane, 1934; Bryson, 1945). The Earth's origins had begun chaotic, perhaps a universal ocean, perhaps molten, certainly uninhabitable (Woodward, 1695: 242f; Catcott, 1761: 26f; Whitehurst, 1778:2f). It had now been organized, temperate, cultivated. The forces of chaos had been harnessed. "The plains become richer, in proportion as the mountains decay" explained Goldsmith (1774: i, 163). God had laboured for six days. Now was the Sabbath on which He rested, the age of Man. All was for the best. Man should not pry too deeply into the rationale of God's ways, but rather embrace them (Goldsmith, 1820: 84-85).

Cosmogony thereby prescribed the terms of the civilization in which Nature made men live. In the 1690s John Ray had God spell it out thus:

> I have placed thee in a spacious and well furnished World. . . . I have provided thee with Materials whereon to exercise and employ thy Art and Strength. . . . I have distinguished the Earth into Hills and Valleys, and Plains and Meadows, and Woods; all these parts capable of Culture and Improvement by thy Industry; I have committed to thee for thy assistance in thy labours of Plowing, and Carrying, and Drawing and Travel the laborious Ox, the patient Ass. [1691: 113-114]

Ray's man counted his blessings:

> I perswade my self, that the bountiful and gracious Author of Mans Being . . . delights in the Beauty of his Creation and is well pleased with the Industry of Man, in adorning the Earth with beautiful Cities, and

> Castles, with pleasant Villages, and Country Houses [and] . . . whatever else differenceth a civil and well-cultivated Region from a barren and desolate wilderness. [Ray, 1691: 484]

The same message—of the opportunities and duties of a life of civilized industry within the established social structure—held good eighty years later in Oliver Goldsmith's cosmogony:

> Such are the delights of the habitation that has been assigned to man. . . . But while many of his wants are thus kindly furnished, on the one hand, there are numberless inconveniences to excite his industry on the other. This habitation, though provided with all the conveniences of air, pasturage, and water, is but a desert place, without human cultivation.
>
> A world thus furnished with advantages on the one side and inconveniences on the other, is the proper abode of reason, is the fittest to exercise the industry of a free and a thinking creature. . . . God beholds, with pleasure, that being which he has made, converting the wretchedness of his natural situation into a theatre of triumph; bringing all the headlong tribes of nature into subjection to his will; and producing that order and uniformity upon earth, of which his own heavenly fabric is so bright an example. [1774: i, 400-1][6]

Of course the telos of earthly existence remained heaven: "God has permitted thousands of natural evils to exist in the world, because it is by their intervention that man is capable of moral evil, and he has permitted that we should be subject to moral evil, that we might do something to deserve eternal happiness by showing that we had rectitude to avoid it" (Goldsmith 1774: i, 20). But the future state—heaven and hell—so physically and temporally important in Stuart cosmology, later became intangible. It was not to distract the gaze from the present. Theorists like Catcott, Whitehurst, Goldsmith, Raspe, and Hutton severed futurity from their discussion of the Earth (cf. Walker, 1964).

Post-1688 theories of the Earth, then, ratified and naturalized the present. They dictated an antibucolic, antiescapist, anti-Utopian, antichiliastic reality principle. The divine order of Nature required men to maximize their individual blessings within the natural and social systems. The natural history of Man in creation legitimated the superiority of the wise over the masses, Europeans over other races, men over women, humans over animals. It explained diversity of race and language, the degeneracy of Negroes, the tenacity of heathenism, the path of the true church (Piggott, 1950). The duty of understanding cosmogony elevated the mind to higher views of things and taught devotion.

CREATION IN CRISIS

Why then did the genre not work—or at least, not last? The question is subtle. For clearly, in so complex a society as that of 18th-century Britain one would not expect so intricate an intellectual structure as a theory of the Earth to be playing—or failing to play—a direct role in maintaining the social structure. Theorists themselves acknowledged the necessarily oblique way their works could speak to society in arguing that the Scriptural cosmogony itself told some truths hieroglyphically to prevent desecration by the vulgar. The problem has many dimensions. One permanent infrastructural weakness was that this particular gospel lacked a tightly disciplined cadre of evangels. No sworn company zealously guarded orthodoxy and monopolized its mysteries. Theories of the Earth were written by utterly heterogeneous individuals—theologians, general scholars, natural historians, popular writers (like Goldsmith), clergy, laymen, some of them scientifically expert, others incompetent; many idiosyncratic and eccentric.[7] Theorists tended to be sui generis rather than the public spokesmen of discrete, well-defined, sectional interests. Diversity spawned conflict and polemic. The theory of the Earth lacked the repetitive platitudinous sonorities so vital to the bland mainstream ideologies which gushed from the Bench, the Latitudinarian pulpit, and the Commons.

For postrevolutionary Britain notably lacked the institutions to shore up scientific orthodoxy. The enfeebled Anglican Church could not address the entire nation, and itself fell into schism over Wesley. Dissent was impotent. The universities declined—Oxford remaining chronically truculent. Unlike the continental absolutisms, there were no centrally directed propagandist academies of learning. The Royal Society dared not, and could not endorse a particular religious, moral, and social orthodoxy. Learning became increasingly fragmented, individual, market-oriented (cf. Plumb, 1973; Saunders, 1964). Hence, whilst most theorists agreed in their broadest commitments, in point of detail most differed *cap à pé* over the hermeneutics of Scripture, or the true mechanism of the Flood, the Diaspora, the chemistry of Creation, the age of the Earth, the longevity of the patriarchs (Egerton, 1966), the location of Eden, and so forth. And because so many theorists cantankerously insisted that theirs was the sole deliverance of science, philosophy, and religion from utter discredit, the anarchic cacophony of opinion deafened ears and invited satire. Thus the demeaning absurdity of Burnet and Warren wrangling whether the Deluge had killed off mankind by drowning or starving; or Catcott's need to ponder whether the Antediluvians might have been sufficiently strong swimmers to escape the Flood (1761: 10).

In part, the kaleidoscopic range of theories also reflects the heterogeneity of political and eccelesiastical affinity and interests in Hanoverian Britain. For while practically all theorists could formulate consonant *aspirations* (that a theory should prove God and His wise Creation and Government of the world), realizing these aspirations was sought along many paths—of Anglican and nonconformist, High and Low Church, philosophe and anti-Enlightenment thinkers, Christians and Deists. Thus Hutchinsonians and liberal Anglicans alike were loyal Hanoverians. Yet for Hutchinsonians the present order of the Earth was primarily a trial, for liberal Anglicans it was reformative (Neve and Porter, 1977). Similarly, both Hutton and de Luc loyally espoused Georgian government, and claimed they located harmonious order in the terrestrial system. Yet their theories were poles apart, and de Luc polemicized against Hutton.

In part, also, the anarchy of contradictory theories sprang from cognitive tensions and overcommitments inherent in the genre. Their key project was to support a divine cosmology by rendering Biblical history and the current terrestrial economy natural and rational. For while stupendous catastrophes could mystify the vulgar (and even at times of crisis, such as the London and Lisbon earthquakes, the intelligentsia [Kendrick, 1956; Rousseau, 1968; Taylor, 1975]), only liberal inquiry into Nature was worthy of independent-minded Englishmen in an age of liberty and Reason.[8] For the appeal to objective Nature was of course attractive: whatever was rational was right. For example, Burnet sought to demonstrate how aging and death necessarily flowed from Adamic sin which had transformed environment and climate. These, once understood, had to be borne, despite all the quackery of "Enthusiasts in Philosophy, as well as in Religion" (1684: 214).

Rendering the status quo natural was a relatively unproblematic ideological task in respect of many sectors of Georgian life. In the higher flights of speculative metaphysics rationalizing struck few mines: cosmic optimism, the great chain of being, mind-body dualism, sufficient reason, and theodicy solutions all commanded broad assent. Likewise the everday flow of society could be scientized in quantified, law-governed, deterministic terms, in accord with capitalist instrumental desire for prediction and managerial control. The century produced a plethora of natural histories, or theories, of education, psychology, morals, the personality, the growth of wealth, law, social relations, and politics. Such Newtonian social sciences rationalized, vindicated, and spurred the new relations of an individualist, consumption-oriented, industrializing society: the mechanisms of the market, the pursuit of profit, utilitarian hedonism, and cost-benefit analysis (cf. Viner, 1972).

Naturalizing the sacred history of Man in Creation, however, posed far greater intellectual ambiguities.[9] In part, these were conflicts of intentions. For most theorists affirmed the core dogmas of revealed Christianity. The essence of these lay beyond the visible world, in the God who had created Nature, and who transcends it by miracles and mysteries (e.g., via holy men, prophecies, providences). The Christian cosmogony hinged on the reality—indeed, the ontologically superior reality—of that supranatural world, since it postulated the very creation of all dimensions of Nature—time, space, and matter—miraculously, ex nihilo, by God. On it hung Christianity's spiritual crux, the attainment of personal immortality for the individual, immaterial soul, through salvation via Jesus Christ. The importance of the doctrine of an afterlife for all shades of 18th-century religion, considered both as personal faith and as public code, needs no elaboration.

Fathoming how to render the supranatural physically demonstrable, without absorbing it, proved—not surprisingly—as elusive as squaring the circle. A minority—Deists—cut the Gordian knot and stipulated that cosmogony could not be mysterious. But most thought this cure was worse than the disease. Denial of the supernatural in cosmogony was the thin end of a wedge which would negate the Church, Mind, the Soul, and finally God himself—and thereby human interests and values, morals and meaning. The *Critical Review* proclaimed (1757: 97) that a natural history of religion was as nonsensical as a moral history of meteors.

Yet few Georgian intellectuals wished to inhabit a world *currently* haunted by spiritual manifestations, miracles, and divine interventions, rubbing shoulders with self-styled saints and prophets. Only the dreaded enthusiasts—the French Prophets, Wesleyans, certain Quakers, and other religious "fanatics"—claimed that the age of miracles had not passed. Such zealots could be gelded by castrating their cosmology. Confronting Wesley, Archbishop Herring argued that the preferable proof of God was the "rising and setting of the sun" rather than "all the extraordinary convulsions of Nature put together" (Nichols, 1812-1816: v, 244). The revolution of the saints had proved too disruptive in the previous century. Polite society no more wished to open the door to philosophical enthusiasts, radicals and Church reformers than to witches, the possessed, Quakers, astrologers, faithhealers, and adepts of the black arts. The world had been turned upside down; it must be kept righted.

Hanoverian ideologues hence prescribed a stable, rational *present* (though one which would allow sufficient to Providence to twit atheists and materialists); but a miraculous and catastrophic *past* for Man on Earth (though one which men of discretion could believe). There was no shortage of mongrel attempts to cobble together such a synthesis of materialism and idealism. Theorists from Woodward to de Luc sought the balance of nature and supernature, of mechanism and miracles, which would satisfy conservative reasonableness and vindicate divine government.

There was, of course, a sliding—not to say, slippery—scale of positions whence to pronounce: Here I stand. For some the urgent task was to make Christianity believable. Thus Burnet was prepared to disregard the Pentateuch, ditch the Hexaemeron, and assert reasonableness as the criterion of Christian cosmogony (1684: 282). The physical mechanism of the Deluge must be as natural as the cracking of dried mud. Similarly, Robert Clayton, Bishop of Clogher, evaporated Noah's flood into a mere partial deluge (1752). A universal deluge was a physical impossibility exactly as a vindictive and arbitrary God was a moral incredibility. Yet the *via media* was delicate. Burnet would still see the preserving of the Ark from shipwreck as a miracle.

Towards the other end of the spectrum, others rejected this "naturalization" of Christianity. Burnet met with a storm of protest when he rationalized Eden, Adam, Eve, the serpent, and original sin (Redwood, 1976: 118-119). The Hutchinsonians stoutly argued for the literal (though emblematic) truth of the Bible, insisting on the irreducibly miraculous turning points of Earth and human history. For God had specifically chosen to transcend Nature, in order to impress upon errant mankind that He—not Nature—was sole divinity. Yet for their part, the Hutchinsonians repudiated resolving the whole of Nature into miracles. That was vulgarly superstitious, and in any case would gain no credence in what they sarcastically called these "enlightened" times (Neve and Porter, 1977).

The dilemma was real. John Ray's work reveals it clearly. He mocked theorists of the Earth, yet in his own *Three Physico-Theological Discourses* (1693) wrote one himself. But there he concluded that science could not understand the great events of Earth history, which hence remained a mystery. Nevertheless, it hardly seems that Ray *wished* scientific method to leave Creation an impregnable mystery: he was genuinely perplexed. A century later the *Monthly Review* was still experiencing the same dilemma. Reviewing Burtin's dissertation on the Deluge, it noted that the author took a wholly literalistic reading of the Scripture, thereby rendering Earth history utterly miraculous. Such a strategy endangered the credit of Moses no less than the assaults of Voltaire, and paved the way for sceptics (1790). This dilemma could be turned to advantage. Thus William Buckland argued that if Creation had presented no conundrums, it would be hard to believe "that this world which we inhabit is the production of that mysterious Being, whose ways are unsearchable, and his works past finding out" (1820: 33). But much more common was Josiah Wedgwood's reproach to John Whitehurst's *Inquiry*, that such clumsy attempts to divinize philosophy and philosophize Moses discredited both (Schofield, 1963: 177).

There is no need to pinpoint here each theory on the Georgian graph of nature and supernature, faith and reason, literalism and liberalism. The point is that there was no resolution, no consensus. Each accused the others of erring too far either—or all—ways. Intellectuals could not agree amongst themselves on the strategy of establishing a transcendental cosmogony upon a rational basis. In having to serve worldly and transcendental purposes the theory of the Earth was overpressurized. Now, other aspects of religion were of course also experiencing tensions. But in many areas these could be practically brushed aside. Soteriological incoherence could be soothed with a bland Arminian injunction to behave. Cracks in the theodicy could be papered over: "Cease then! Nor Order, Imperfection name!" But cosmogony could not— initially, at least—be left in decent obscurity. For it told a unique and critical series of events in the constitution of the world, about which Holy Writ itself was so inescapably explicit. Eighteenth-century naturalists had to grapple with the most comprehensive cosmogony of any of the world's great religions, planted firmly at the opening of its Sacred Book, a cosmogony full of obscurities such as its abysses, firmaments resting on pillars, waters above the firmament, and geocentrism (Grierson, 1975; Jacobs, 1975). In short, although it was attractive for Georgian cosmology to use Nature both to underpin current society and to signpost Eternity, *theorizing* the relations between Creation and coal mines, or Adam and bourgeois accumulation, remained problematic.

The consequences were manifold. Instead of propounding a resonant, unison chorus of cosmogony suitable for indoctrination, the period was punctuated by acrimonious sniping amongst those who were purporting to define Scripture, Truth, and the cosmic order. Controversy in the reigns of William and Anne between Burnet, Woodward, Whiston, Arbuthnot, Keill, Harris, and others produced only gall. And dyspepsia bubbled through the century. The leading reviews treated new theories with scorn. Amongst Christians, Hutchinson slated the Newtonians, Catcott lambasted Bishop Clayton, Kirwan and de Luc squabbled amongst themselves. But no more could non-Christians agree with each other. Hutton diverged from Buffon, Toulmin from Hutton, and Erasmus Darwin from all of them. No consensus was near to being achieved. Deists had little to worry about from disagreement. Being a hybrid of Young Turks and *enfants terribles,* they possessed power without responsibility. But rifts within the Christian camp were far more wounding. In a plaintive letter to Dr Parsons in 1745, William Borlase lamented that theoretical controversies among Christians were threatening to discredit Moses (Nichols, 1812-1816: v, 301).

The second consequence is that the theory of the Earth began to shiver into fragments. Stuart theorists had aimed (witness Burnet's pictorial frontispiece: 1684) to articulate the comprehensive theory of Man from Creation to Consummation. Christianity being all-embracing and monotheistic, it seemed vital for its credit that the theory of the cosmos should remain unitary. Such attempts had been staked out by Raleigh (1614) and Grew (1701), by Burnet (1684) and Whiston (1696). But later accounts became more episodic. Catcott tackled the diluvial history of Man. Whitehurst discussed only Man's longevity. John Williams concentrated on the dispersal of Man after the Flood, showing how it had been the Welsh who had populated America (1789). Many theorists, such as James Douglas (1785) and de Luc (1790-1791, 1793-1774) fell quiet about the precise early sacred history of man, though concerned still with the great hinges of his destiny. The *future* of the globe disappeared almost entirely from such theories.

A potentially unitary science of Man was meanwhile being challenged in another way. The new Enlightenment social sciences—economics, demography, linguistics, political science—sought a philosophical (and frequently critical) interpretation of Man's nature by speculatively probing his origins. The "state of Nature" had been popularized, e.g., by Locke, to quash Stuart patriarchalism. Now it was proving a Trojan horse. Conjectural histories were not easy to square with Christian genealogy, and could readily be used by opponents such as the *philosophes* as an alternative aetiology.

Furthermore, the material surge of European imperialism and its accompanying exploration and scientific research were revealing a terraqueous globe—the world of Buffon, Cook, and Humboldt—strikingly different from that of the patristic writers. How was New World fauna assembled for, and then squeezed into the Ark?—the size of the Ark, which Stuart scholars debated with relish, was swept under the carpet in the next century. Aligning evidence of race, language, strata, fossils, and human remains with the biblical six thousand years became a conundrum. Adam, though not yet dead, was talked about less in polite circles (Greene, 1959).

The consequence was a growing diffidence in articulating in natural, empirical, "falsifiable" form certain features of the cosmogony. Following Boyle's injunction that God was "so far from being unwilling that we should pry into His works," e.g., by beginning "the Book of Scripture with the description of the Book of Nature," that "He imposes on us little less than a necessity of studying them" (1744:ii, 19), Stuart theorists had felt justifiable confidence in juxtaposing the written and natural books of God's truth. A century later, however, they had become far more leery about offering hostages to fortune. It was to become a popular line of disengagement that Scripture was not intended to teach natural history; Scripture taught moral truths not natural.

Reviewing Philip Howard's theory, the *Analytical Review* struck a new note (1797:238). Man was necessarily concerned about his ancestry. But the "various systems. . .framed" have been "so obscure in themselves, and contradictory to each other, as to afford the cautious inquirer little satisfaction, and indeed scarcely to do more than teach him an humiliating lesson on the imbecility of the human understanding." From a more conventionally pious standpoint, Richard Joseph Sulivan reached a no less depressing conclusion. All hitherto existing philosophies of creation, he wrote, were proved "contradicted by reason and experience" (1794:i, 98). Sulivan urged great caution. "The mystery of creation is greatly beyond the powers of our intellect" (1794:i, 95). All that theorists have "imagined about the chaos and the formation of the universe, has been found baseless, and unintelligible" (1794:i, 97-98). The more conservative Anglican *Critical Review* delivered a similar verdict on Robert Miln's *Course of Physico-Theological Lectures.* Noting the general dangers of natural theology —converting "blemishes into blessings" (1788:35)—it found that when Miln rationalized the Creation and the Deluge

> he is very unequal, and very inconsistent. He contends, for instance, for the philosophical accuracy of every part of the Mosaic account of the creation. . .while in another place, where the motion of the sun is mentioned, he expressly tells us, that the scriptures are not designed to teach us philosophy.

The review concludes, ominously: "It would be improper to consider the history of the Creation as related by Moses, too minutely" (1788:35).

New strategies therefore needed to be devised. On the one hand, mounting criticism was directed against minutely detailed Mosaic cosmogony. Christians did not *wish* to budge one jot from their faith in man's novity and special place in Creation—such beliefs were held tenaciously till late Victorian times. Rather the fear was that penetrating the veil had become desperately counterproductive. Hence, attempts were made to expound a Christian history avoiding meticulous concentration on the problem areas. John Walker, Natural History professor at Edinburgh and a leading Scottish churchman, programmatically announced he "would not wish to be thought to deliver any thing like Theory, but merely a natural history of the earth" (1966: 180)—though he felt confident enough to draw such explicit social conclusions from the natural order as that by God's hand "the Many submit to be governed by the Few" (Occasional Remarks).

TRANSFORMATION

The 1770s and 1780s thus saw a lower-profile, slimmed-down theory of the Earth amongst such naturalists as de Luc, Whitehurst, and James Douglas on the Christian side. These were perhaps paralleled by the theories of the Deist Hutton, who likewise left man's natural history cautiously understated. Such trends, however, were overtaken by the French Revolution. The classic Burkean indictment against the Revolution attributed its origins to the abstract theoretical speculations of the atheistic philosophes. Voltaire, d'Holbach, Diderot, Helvétius—it was well known—had produced their own dogmatically anti-Christian constructions of the history of Nature and Man (Manuel, 1959). These provoked some renewed point-by-point rebuttals in the cause of Genesis literalism (cf. de Luc, 1790-1791, 1793-1794; Howard, 1797), as well as hysterical condemnation of such theories as utterly destructive of society—thus John Williams damned James Hutton's ideas as inevitably triggering "anarchy, confusion and misery" (1789:vol. 1, lx).

The chief response, however, was that the theory of the Earth had been so thoroughly contaminated—or even appropriated—by the enemy that Christians must abandon the genre altogether as too dangerous. Chastened by the late events in France, the Reverend George Graydon stated that theories were at best

> to be considered as contributing but remotely to the more useful and serious objects of life. But when applied, as we know they have been too

> often, to excite and diffuse doubts of the most essential truths, and ultimately to sap the foundations of religion, and with it, of both private and public virtue, order and happiness, and indeed of the very existence of civil society, as too fatal modern experience has shown, it is not easy to say whether we shall be most struck with the vanity and presumption, the folly, or the wickedness of the attempt. [1791:311]

Repudiating cosmogonies, the conspirator-finder general, John Robison, tendentiously argued that Man's destiny should not be thought to hinge upon the constitution of Nature: "It is amusing to observe the earnestness with which they [i.e., Jacobin radicals] recommend the study of natural history. One does not readily see the connection of this with their ostensible object, the happiness of man" (1797:543). This was the death-knell of the classic theory of the Earth. For at this point theologians and men of science began to negotiate a trial separation. Early 19th-century churchmen began the trek away from a rationalizing, naturalizing theory—the evidences of Christianity—back to a religion of unrepentent mystery: theological obscurantism mirroring political.[10] At the same time the scientific elite altogether repudiated cosmogony and defined its newly crystallizing science of geology over and against it in the manner of chemistry to alchemy, or astronomy to astrology. John Kidd announced the evidence for a theory of the Earth was imperfect (1815), and John Playfair dismissed theories as an "unreal mockery," "a species of mental derangement" (1811:207-208).

By the end of the 18th century something like a coherent community of expert natural philosophers investigating the Earth had gelled out of the previous confusion earlier described. Its characteristics were that it was lay, rather than clerical, comprising middle-class, professional men of science and gentlemen amateurs, rather than professorial academics. It included men of all religious persuasions. Such a group publicly championed the autonomy and integrity of their pursuit of geology by scientific methods, while more privately maintaining a deep commitment to the spiritual superiority of Man, grasped within a general religious framework. Living in revolutionary times, they were only too well aware that the very activity of geology was liable to suspicion. Hence, they developed a two-fold strategy for coping with the pressures of social control being exerted *over themselves* and for projecting their own values in the most favourable light. On the one hand they advanced an extremely positivistic, Baconian methodology. By describing and mapping the physical configuration of the strata, geology would endorse reality. It would be a science with its boots tramping the ground, rather than its head in the clouds. Thereby geology could unite its own practitioners to present a consolidated front to the world in institutions like the Geological Society of London.

But the strategy was also to cordon Man off from the history and physical system of the Earth. Replacing the traditional goal of using Earth science to chart the course of God's dealings with Man, early 19th-century geology specialized in technical stratigraphy, palaeontology, and mineralogy. Creation was side-stepped—it had become a separate subject, Cosmogony, with which geology programmatically would have nothing to do, being concerned only with the present system of the Earth (cf. Lyell, 1830-1833:i, 1; Porter, 1976; Kidd, 1815). Geologists from William Smith to Geikie kept quiet on the Earth's ultimate origins and on the origin of species (Whewell said science must look piously upwards). Very few among the charmed inner circle of the Geological Society discussed the Deluge. When they did—like Buckland—they were rebuked by their fellows, or soon themselves recanted, as did Sedgwick (Porter, 1977:208f). Guardianship of the traditional, fully articulated theory of the Earth became the preserve of the "scriptural geologists," marginal figures who abused and were abused by the new leaders of the geological community.[11] Elite geologists by contrast contented themselves with platitudinous pieties and Romantic aesthetics.

What had happened was not of course that geological science had superseded religion, ideology, or interests. As Gillispie admirably showed, 19th-century geology continued to carry an ideological load heavy enough to stagger Atlas (1951). Lyell, Sedgwick, Whewell, Scrope, and Murchison all believed, no less than Woodward or Hutton had, that science had a mission in the creation of the new society, and their several views of geology just as clearly encoded social preferences (Porter, 1976; Rudwick, 1974). Rather, in the new configuration of geological knowledge discussion of Man had become formally separated from that of Nature. Man was fenced off from geological history (Bynum, 1976). In Lyell's geology, Man was ontologically distinct from the rest of Earth's history, in being moral, spiritual, and progressively rational. Man alone was recent.

Thus for British palaeontologists of the first half of the 19th century it was axiomatic that human remains would not be found alongside the bones of extinct creatures, because it was unthinkable that Man did not constitute an epoch of his own, independent of Earth history. Hence the extraordinary paradox of Buckland, the vindicator of the Deluge, consistently refusing to "discover" human *reliquiae diluvianae.* Man had become reconceptualized as the culmination of the eons-long development of the Earth, the final cause of all change, but essentially an object to be studied by social anthropologists, archaeologists, antiquarians, philologists, folk-lorists, and other nonnaturalistic cultural historians. Of course, conceptions of the place of Man in and out of Nature continued to shape thoughts about the terraqueous globe; and dilemmas intensified. But these needed to be discussed in the private agony of notebooks, as by Lyell (1970). Burnet's ambition of a "moral or philosophick history of the world" had become too painful to explicate.

In this chapter I have tried to show that naturalistic cosmogonies were a form of knowledge arising out of the need to provide new foundations for a conception of Man in the cosmos at a time of social, political, and intellectual change when traditional religious dogmas had ceased to command unanimity. The theory of the Earth itself, however, achieved little stability. For Georgian values were themselves torn between the goals of a secularizing Enlightenment "materialism"—the endorsement of the here-and-now—and the "idealism" of Christian salvationism. The ideological guardians were split between clerical and lay elites. Ideological projections sundered on tensions between the desire to affirm the day-to-day world, and to guarantee transcendental values. The alliance was uneasy between otherworldly mythology, revealed in an authoritative Holy Writ, and the need to create a usable, pragmatic, human, and natural past (Chadwick, 1975; Glasner, 1977). The dying cosmogonical bird was swept off the stage, hustled by the French Revolution. It became expedient to construct a new cognitive map, in which natural science and human values could flourish precisely by tactfully being kept formally more separate rather than integrated. The theory of the Earth died, but the interests which produced cosmic ideologies lived on. *In nova fert animus mutatas ducere formas.*

NOTES

1. Theories of the Earth were a European phenomenon. In this chapter, however, I shall deal only with British instances.

2. Throughout this chapter I shall mean by "traditional cosmogony" that very complex interweaving of Middle Eastern, Greek, Roman, Judaic, and Christian elements which had been received and developed by Medieval and early modern Western European society. There is clearly no room here to elaborate on the functions of that cosmogony, but for a comparative perspective see Needham (1959), Blacker and Loewe (1975), and Douglas (1970). Equally, there is no room here to investigate the *psychological* meaning of this cosmogony (the Earth as Mother, and so on). For stimulating introductions see Eliade (1969), Kolodny (1975), and Tuan (1974).

3. For a particularly illuminating question-and-answer exposition of the heart of this cosmogony see Nicholls (1696).

4. Classic studies attempting to demonstrate this Newtonian hegemony are Schofield (1970), Thackray (1970), Heimann and McGuire (1971). More recent historiography has become sceptical as to how coherent such a Newtonian tradition actually was (cf. Home, 1977; Cantor, 1978). Nevertheless, Newtonian prestige was never seriously threatened.

5. Though of course this is a highly complex issue because Burnet himself rejected "literal" accounts of the drama of original sin.

6. To say that Goldsmith urged individual reconciliation to the actual world of political, economic, and familial structures is not the same as to say that he was an ideologue of "capitalism." He was writing "The Deserted Village" (1770) at the same time as his *History of the Earth.*

7. Woodward, for instance, was so abusive he was expelled from the council of the Royal Society. Whiston gave itinerant lectures on the end of the world. Da Costa and Raspe were both punished for embezzlement. Hutchinson and Catcott were much satirized. Kirwan became a pronounced eccentric, and so on.

8. For contemporary opinion, see, for example, Sulivan (1794:ii, 237). For Reason in 18th-century theology see Pattison (1859) and Cragg (1964).

9. The best parallel theological controversy is the Trinitarian-Arian-Socinian, since the same issues were at stake (could there be a natural understanding of Jesus?).

10. This led ultimately to Newman's strategic dualism: "Theology and Physics cannot touch each other" (Cannon, 1964:495). For literary attempts to render annals of ancestry into new mythical form, see Shaffer (1975).

11. Sharon Turner was one such scriptural geologist who believed that it was Judas-like treason—and quite unnecessary—for Christians to abandon a minutely detailed, scientifically attested explicit cosmogony (1836:i, viii).

REFERENCES

ALLEN, D.C. (1949) The Legend of Noah. University of Illinois Studies in Language and Literature, Vol. 33. Urbana, Illinois.

Analytical Review (1797) "Review of P. Howard, The Scriptural History of the Earth and of Mankind." 25:238-247.

AULT, D.D. (1974) Visionary Physics: Blake's Response to Newton. London: Univ. of Chicago Press.

BAILEY, E.B. (1967) James Hutton, the Founder of Modern Geology. Amsterdam: Elsevier.

BAKER, H.C. (1952) The Wars of Truth. Cambridge, Mass.: Harvard Univ. Press.

BECKER, C. (1967) The Heavenly City of the Eighteenth-Century Philosophers. London: Yale Univ. Press.

BLACKER, C. and LOEWE, M. [eds.] (1975) Ancient Cosmologies. London: George Allen & Unwin.

BOYLE, R. (1744) Works, to which is prefixed the Life of the Author (by T. Birch). 5 vols.London.

BRYSON, G. (1945) Man and Society: The Scottish Inquiry of the Eighteenth Century. Princeton, N.J.: Princeton Univ. Press

BUCHDAHL, G. (1961) The Image of Newton and Locke in the Age of Reason. London: Sheed & Ward.

BUCKLAND, W. (1820) Vindiciae Geologicae: Or, the Connexion of Geology with Religion Explained. Oxford.

BURNET, T. (1684) The Theory of the Earth (trans. of 1681 Latin orig.). London. (Also ed. B. Willey, 1965, London: Centaur Press.)

BYNUM, W.F. (1976) "The Blind Man and the Elephant: Toward a History of Pre-History." Unpublished paper delivered to the Conference on New Perspectives in the History of the Life Sciences, Cambridge, England.

CANNON, W.F. (1964) "The normative role of science in early Victorian thought." J. of the History of Ideas 25:487-502.

CANTOR, G.N. (1978) "The historiography of 'Georgian' optics." History of Science, 16:1-21.

CAPP, B. (1972) The Fifth Monarchy Men. London: Faber & Faber.

CATCOTT, A. (1761) A Treatise on the Deluge. London.

CHADWICK, W.O. (1975) The Secularization of the European Mind in the Nineteenth Century. Cambridge: Cambridge Univ. Press.

CLAYTON, R., Bishop of Clogher (1752) A Vindication of the Histories of the Old and New Testament. Dublin.

COLLIER, K.B. (1934) Cosmogonies of Our Fathers. New York: Columbia Studies in History.

CRAGG, G.R. (1964) Reason and Authority in the Eighteenth Century. Cambridge: Cambridge Univ. Press.

———(1950) From Puritanism to the Age of Reason. Cambridge: Cambridge Univ. Press.

CRANE, R.S. (1933-1934) "Anglican apologetics and the idea of progress, 1699-1745." Modern Philology 31:273-306, 349-382.

Critical Review (1788) "Review of Robert Miln, A Course of Physico-Theological Lectures." 65:35-40.

———(1757) "Review of D. Hume, Four Dissertations. 1. Natural History of Religion." 3:97-107.

DAVIES, G.L. (1966) "The eighteenth century denudation dilemma and the Huttonian theory of the earth." Annals of Sci. 22:129-138.

DOUGLAS, J. (1785) A Dissertation of the Antiquity of the Earth. London.

DOUGLAS, M. (1970) Natural Symbols: Explorations in Cosmology. London: Cresset Press.

EGERTON, F., III (1966) "The longevity of the Patriarchs, a topic in the history of demography." J. of the History of Ideas 27:575-584.

ELIADE, M. (1961) Images and Symbols. London: Harvill Press.

GARFINKLE, N. (1955) "Science and religion in England 1790-1800: The critical response to the work of Erasmus Darwin." J. of the History of Ideas 16:376-388.

GAY, P.J. (1967-1970) The Enlightenment: An Interpretation. London: Weidenfeld & Nicolson.

GILLISPIE, C.C. (1951) Genesis and Geology. Cambridge, Mass.: Harvard Univ. Press.

GLACKEN. C.J. (1967) Traces on the Rhodian Shore. Berkeley: Univ. of California Press.

GLASNER, P.E. (1977) The Sociology of Secularisation. London: Routledge & Kegan Paul.

GOLDSMITH, O. (1774) An History of the Earth and Animated Nature. London.

———(1766) The Vicar of Wakefield: A Tale. Dublin.

GRAYDON, G. (1791) "On the fish enclosed in stone of Monte Bolca." Transactions of the Royal Irish Academy 5:281-317.

GREENE, J.C. (1959) The Death of Adam. Ames: Iowa State Univ. Press.

GREW, N. (1701) Cosmologia Sacra. London.

GRIERSON, P. (1975) "The European heritage," pp. 225-258 in C. Blacker and M. Loewe (eds.) Ancient Cosmologies. London.

GUERLAC, H. (1965) "Where the statue stood: Divergent loyalties to Newton in the eighteenth century," pp. 317-334 in E. R. Wasserman (ed.) Aspects of the Eighteenth Century. London: Oxford Univ. Press.

HALE, M. (1677) The Primitive Origination of Mankind. London.

HARRIS, J. (1697) Remarks on Some Late Papers Relating to the Universal Deluge and to the Natural History of the Earth. London.

HAY, D. (1975) "Property, authority and the criminal law," pp. 17-64 in D. Hay, P. Linebaugh and E.P. Thompson (eds.) Albion's Fatal Tree. London: Allen Lane.

HAZARD, P.G.M.C. (1964) The European Mind, 1680-1715 (trans. by J.L. May). Harmondsworth: Penguin.

HEIMANN, P.M. and McGUIRE, J.E. (1971) "Newtonian forces and Lockean powers. Concepts of matter in eighteenth century thought." Historical Studies in the Physical Sciences 3:233-306.

HILL, C. (1977) Milton and the English Revolution. London: Faber.
_____(1971) Anti-Christ in Seventeenth Century England. London: Univ. of Newcastle-upon-Tyne Publications.
HOME, R. (1977) "Newtonianism and the theory of the magnet." History of Science 15:252-266.
HOOKE, R. (1705) The Posthumous Works of Robert Hooke. London: R. Waller.
HOWARD, P. (1797) The Scriptural History of the Earth and of Mankind. London.
HUTTON, J. (1795) The Theory of the Earth. Edinburgh.
JACOB, M.C. (1977) "Newtonianism and the origins of the Enlightenment." Eighteenth Century Studies 11:1-25.
_____(1976) The Newtonians and the English Revolution 1689-1720. Hassocks: Harvester Press.
_____and LOCKWOOD, W.A. (1972) "Political millenarianism and Burnet's *Sacred Theory*." Science Studies 2:265-279.
JACOBS, L. (1975) "Jewish cosmology," pp. 66-86 in C. Blacker and M. Loewe (eds.) Ancient Cosmologies. London.
JONES, R.F. (1936) Ancients and Moderns. St. Louis: Washington University Studies, n.s., Language and Literature, 6.
KEILL, J. (1698) An Examination of Dr. Burnet's Theory of the Earth, Together with Some Remarks on Mr. Whiston's New Theory of the Earth. Oxford.
KENDRICK, T.D. (1956) The Lisbon Earthquake. London: Methuen.
KIDD, J. (1815) A Geological Essay on the Imperfect Evidence in Support of a Theory of the Earth. Oxford.
KIERNAN, C.P. (1973) The Enlightenment and Science in Eighteenth Century France. Banbury: Voltaire Foundation.
KOLODNY, A. (1975) The Lay of the Land. Metaphor as Experience and History in American Life and Letters. Chapel Hill: Univ. of North Carolina Press.
KUBRIN, D.C. (1968) "Providence and the mechanical philosophy. The creation and dissolution of the world in Newtonian thought. A study of the relations of science and religion in seventeenth century England." Ph.D. dissertation, Cornell University.
KUHN, A. (1961) "Glory or gravity: Hutchinson vs Newton." J. of the History of Ideas 22:303-322.
LAMONT. W.M. (1969) Godly Rule. London: St. Martin's Press.
LASLETT, T.P.R. (1949) Patriarcha, and other Political Writings of Sir Robert Filmer. Oxford: Basil Blackwell.
LOVEJOY, A.O. (1936) The Great Chain of Being. Cambridge, Mass.: Harvard Univ. Press.
LOWENTHAL, D. (1975) "Past time, present place: Landscape and memory." Geographical Rev. 65:1-36.
LUC, J.A. de (1793-1794) "Geological letters addressed to Professor Blumenbach," British Critic 2:231-239, 351-358; 3:110-118, 226-237, 467-478, 589-598.
_____(1790-1791) "Letters to Dr. James Hutton, F.R.S., Edinburgh, on His Theory of the Earth." Monthly Rev. 2:206-227, 582-601; 3:573-586; 5:564-585.
LYELL, C. (1970) Sir Charles Lyell's Scientific Journals on the Species Question (L.G. Wilson, ed.). New Haven: Yale Univ. Press.
_____(1830-1833) Principles of Geology. London
MacPHERSON, D.B. (1962) The Political Theory of Possessive Individualism. Oxford: Oxford Univ. Press.

MANUEL, F.E. (1959) The Eighteenth Century Confronts the Gods. Cambridge, Mass.: Harvard Univ. Press.

MARTIN, B. (1735) The Philosophical Grammar. London.

MILLHAUSER, M. (1954) "The scriptural geologists." Osiris 11:65-86.

MILTON, J. (1667) Paradise Lost. London.

Monthly Review (1790) "Review of F.X. Burtin, Réponse à la Question Physique Proposée par la Société de Teyler." n.s. 3:539-544.

NEEDHAM, N.J.T.M. (1959) Science and Civilisation in China. Cambridge: Cambridge Univ. Press.

NEVE, M. and PORTER, R. (1977) "Alexander Catcott: Glory and geology." British J. for the History of Sci. 10:47-70.

NICHOLLS, W. (1696) A conference with a Theist, wherein are Shewn the Absurdities in the Pretended Eternity of the World. London.

NICHOLS, J. (1812-1816) Literary Anecdotes of the Eighteenth Century. London.

——(1809) [ed.] Letters on Various Subjects. . . to and from William Nicolson. London.

NICOLSON, M.H. (1959) Mountain Gloom and Mountain Glory. Ithaca: Cornell Univ. Press.

NOKES, D.L. (1975) "A study of the Scriblerus Club." Ph.D. dissertation, Cambridge University.

PASSMORE, J. (1974) Man's Responsibility for Nature. London: Duckworth.

PATTISON, M. (1860) "Tendencies of religious thought in England, 1688-1750," pp. 254-329 in Essays and Reviews. London.

PHILLIPS, W. (1815) An Outline of Mineralogy and Geology. London.

PIGGOTT, S. (1950) William Stukeley. Oxford: Oxford Univ. Press.

PLAYFAIR, J. (1811) "Review of Transactions of the Geological Society." Edinburgh Rev. 19:207-229.

PLUMB, J.H. (1973) The Commercialization of Leisure in Eighteenth Century England. Reading: Reading Univ. Press.

——(1972) "The public, literature and the arts in the eighteenth century." Pp. 27-48 in P. Fritz and D. Williams (eds.) The Triumph of Culture. Toronto: A.M. Hakkert.

——(1967) The Growth of Political Stability in England, 1675-1725. London: Macmillan.

PORTER, R. (1978a) "Philosophy and politics of a geologist: G.H. Toulmin (1754-1817)." J. of the History of Ideas. 39:435-450.

——(1978b) "George Hoggart Toulmin and man's place in the development of British geology." Annals of Sci. 35:339-352.

——(1977) The Making of Geology: Earth Science in Britain 1660-1815, Cambridge: Cambridge University Press.

——(1976) "Charles Lyell and the principles of the history of geology." British J. for the History of Sci. 9:91-103.

RALEIGH, W. (1614) The Historie of the World. London.

RAY, J. (1693) Three Physico-Theological Discourses. London.

——(1691) The Wisdom of God Manifested in the Works of the Creation. London.

REDWOOD, J. (1976) Reason, Ridicule and Religion: The Age of Enlightenment in England, 1660-1750. London: Thames & Hudson.

ROBISON, J. (1797) Proofs of a Conspiracy Against All the Religions and Governments of Europe. London.

ROGER, J. (1973-1974) "La théorie de la terre au XVIIIe siècle." Revue d'Histoire des Sciences 24:23-48.

ROUSSEAU, G.S. (1968) "The London earthquakes of 1750." Cahiers d'Histoire Mondiale 11:436-451.
RUDWICK, M.J.S. (1974) "Poulett Scrope on the volcanoes of Auvergne." British J. for the History of Sci. 7:205-242.
SAUNDERS, J.W. (1964) The Profession of English Letters. London: Routledge & Kegan Paul.
SCHOCHET, G.J. (1975) Patriarchalism in Political Thought. Oxford: Basil Blackwell.
SCHOFIELD, R.E. (1970) Mechanism and Materialism. Princeton: Princeton Univ. Press.
_____(1963) The Lunar Society of Birmingham. Oxford: Oxford Univ. Press.
SHAFFER, E.S. (1975) "Kubla Khan" and the Fall of Jerusalem. Cambridge: Cambridge Univ. Press.
SHAPIN, S. and BARNES, B. (1977) "Science, nature and control: Interpreting mechanics' institutes." Social Studies of Sci. 7:31-74.
STILLINGFLEET, E. (1662) Origines Sacrae. London
STROMBERG, R.L. (1954) Religious Liberalism in Eighteenth Century England. Oxford: Oxford Univ. Press.
SULIVAN, R.J. (1794) A View of Nature. London.
TAYLOR, J. (1975) "Eighteenth century earthquake theories: A case history investigation into the character of the study of the earth in the Enlightenment." Ph.D. dissertation, University of Oklahoma.
THACKRAY. A.W. (1970) Atoms and Powers. Cambridge, Mass.: Harvard Univ. Press.
TOON, P. (1970) Puritans, the Millennium and the Future of Israel. Cambridge: Cambridge Univ. Press.
TUAN, Y. (1974) Topophilia. Englewood Cliffs, N.J.: Prentice-Hall.
_____(1968) The Hydrologic Cycle and the Wisdom of God. Toronto: Univ. of Toronto Press.
TURNER, S. (1836) The Sacred History of the World. London.
TUVESON, E.L. (1950) "Swift and the world-makers." J. of the History of Ideas 11:54-74.
_____(1949) Millennium and Utopia. Berkeley: Univ. of California Press.
VINER, J. (1972) The Role of Providence in the Social Order. Princeton: Princeton Univ. Press.
WALKER, D.P. (1964) The Decline of Hell. London: Routledge & Kegan Paul.
WALKER, J. (n.d.) "Occasional Remarks" (MS in Edinburgh University Library, DC 2 40).
_____(1966) John Walker: Lectures on Geology, including Hydrography, Mineralogy and Meterology (H.W. Scott, ed.). Chicago: Univ. of Chicago Press.
WHISTON, W. (1696) A New Theory of the Earth. London.
WHITEHURST, J. (1778) An Inquiry into the Original State and Formation of the Earth. London.
WILLEY, B. (1940) The Eighteenth Century Background. London: Chatto & Windus.
_____(1934) The Seventeenth Century Background. London: Chatto & Windus.
WILLIAMS, A. (1948) The Common Expositor: An Account of the Commentaries on Genesis, 1527-1633. Chapel Hill, NC: Univ. of North Carolina Press.
WILLIAMS, J. (1789) The Natural History of the Mineral Kingdom. Edinburgh.
WOODWARD, J. (1695) An Essay Toward a Natural History of the Earth. London.

5

DARWIN AND SOCIAL DARWINISM: PURITY AND HISTORY

Steven Shapin
Barry Barnes

Over the past two decades historians of science have subjected Darwin and his work to concentrated study. There is every reason to think that this interest is increasing and is capable of sustaining a prosperous "Darwin industry" (Ruse, 1974; Loewenberg, 1965; Greene, 1975). In recent years our knowledge of Charles Darwin's own career, publications, and manuscripts has become vastly more detailed and reliable. Since approximately 1960 the focus of research interest has moved from the exegesis of Darwin's main published texts to the minute study of marginalia, notebooks, unpublished sketches, and letters. The main aim of this enterprise has been the reconstruction of the thought processes of a creative genius.

Against the background of a flourishing Darwin enterprise, the relative poverty of the literature on the nature of social Darwinism becomes even more interesting, especially so since many modern studies have also briefly considered the question of "links" between Darwin and social Darwinism. Hofstadter's brilliant study of American social Darwinism is now thirty-five years old, and, apart from Burrow's (1966) anthropologically focused account, there is no

AUTHORS' NOTE: The authors are indebted to their colleague David Bloor for criticisms of an earlier version of this paper.

serious treatment of social Darwinism in Darwin's own country (but see Durant, 1977).

One reason for this is that social Darwinism is a peculiarly intractable subject. It straddles the currently accepted boundaries of academic disciplines; and, appearing to be neither fish nor fowl to both social historians and historians of science, is rarely eaten by either. What, after all, *is* social Darwinism? In the past, historians tended to conceive of it mainly as the naturalistic legitimation of individualistic capitalism and rampant imperialism. More recently, Himmelfarb has characterized it in the "traditional" sense as "a philosophy exalting competition, power, and violence over convention, ethics, and religion," yet also recognized that "In the spectrum of opinion that went under the name of social Darwinism almost every variety of belief was included" (1968:416, 431). Generally speaking, the more modern the study, the more acute has been the realization that social Darwinism had a protean form (e.g., Young, 1969:130-134; Smith, 1972; Bowler, 1978). This literature is replete with references to the wide spectrum of social uses to which Darwinism was put, on "right" and "left" wing; from Sumner, Carnegie, and Theodore Roosevelt to Marx, Henry George, and Kropotkin. Gillispie (1960:342) tells us that "rugged individualists, Marxists, militarists, [and] racists" all found "ideological comfort" in a Darwinian vision. And Robert Young (1971b: 185) notices that "evolution had been invoked to support all sorts of political and ideological positions from the most reactionary to the most progressive."

A closely related problem arises out of recent realizations that social Darwinism was part of a much wider cultural movement. It is argued that social Darwinism was a subspecies of "scientific naturalism"—the tendency increasingly to account for natural phenomena (including man) in nonanthropomorphic, secular, uniform, nonteleological, and materialist terms (e.g., Burrow, 1966; Young, 1970; Turner, 1974). Darwinism and social Darwinism were strands of a *movement*—a movement which included in its leadership, according to one account, Huxley, Tyndall, Spencer, Galton, and W.K. Clifford (Turner, 1974:9). Young's (1970:14) analysis of "The impact of Darwin on conventional thought" likewise makes it clear that

> Darwin's theory and its reception were part of a much larger debate on evolution itself, and more generally on a naturalistic or scientific approach to the earth, life, man, his mind, and society.

Like Turner, Young is convinced that Darwin was one of a leadership cadre, including "Spencer, Wallace, Huxley, [the authors of] *Essays and Reviews*, and John Tyndall" whose work

> was part of a larger movement embracing a number of naturalistic approaches to the earth, life, and man—in utilitarianism, in population theory, in geology, phrenology, psychology, and in theology itself. [Young, 1970:16]

This perspective, undoubtedly the correct one, makes social Darwinism at once more intriguing and more intractable as a subject for empirical enquiry; a "cosmology" is not considered to be the sort of thing one can "document" according to traditional historical procedures.

Nevertheless, the fact remains that social Darwinism has been poorly characterized and studied. What is more, this fact is not convincingly or sufficiently accounted for by technical problems. The fundamental explanation is that there has been little interest in social Darwinism *as such*. Most of whatever historical interest exists has continued to focus on Charles Darwin's relationship to the movement. The central question seems to have been whether the author of the *Origin of Species* of 1859 was in any way either a "part of," or "responsible for", social Darwinism. The generally agreed-upon answer seems to be that the accused was innocent on both counts. This chapter reports on the trial from outside, and reflects upon the likely reasons for, and consequences of, such a protracted and expensive litigation.[1] First we shall review recent literature which has focused upon Darwin's "responsibility" for social Darwinism. We consider the historiographic preoccupations and practices displayed in this literature. In particular, we shall comment on its concern with Darwin's inner mental processes, and conclude by drawing attention to the practical problems this concern raises for historical enquiry.

WAS DARWIN GUILTY?

The defence in the Darwin case has rested upon three assertions. The first is that of internal purity: Darwin's *intentions* and *motives* in writing the *Origin* were above reproach, and his personal *beliefs* in 1859 were innocent of "ideological" taint. The second is purity of ancestry: "influences" upon the *Origin* were entirely wholesome and reputable; in particular, nothing "ideological" was gleaned from Malthus. The third assertion is purity of germ-plasm: nothing untoward could *properly* be deduced from the theory in the *Origin*; truth does not blend with error; insofar as truth was used to justify social Darwinism, it was misused. All the assertions are justified by means of textual exegesis: appropriate glosses are put upon Darwin's writings as indicators of internal psychological states, or as manifestations of appropriate connections between ideas.

Interestingly, some of the most unambiguous testimonials of Darwin's internal purity come from the historian most concerned to depict Darwin's evolutionary theory as embedded in an ideological context. Thus, Robert Young (1971a:443) distinguishes Darwin from other, ideologically motivated evolutionary writers:

> Of course, Darwin's theory was based on a more plausible scientific hypothesis than those of the other evolutionists, and he was much less interested in philosophical, theological, and social issues: he was primarily a naturalist.

Darwin thus contrasts instructively with evolutionists like Spencer and Wallace who were "quite explicit about the role of ideology" in their evolutionary thinking (Young, 1971b:184).And there is only a hint of a shift of emphasis in Young's most recent statement:

> It should be granted that the work and influence of Lyell and Darwin were less intentionally and obviously an expression of more basic socio-economic forces and structures than, for example, the work and influence of Chambers, Spencer and Wallace. [1973:385; but cf. possibly contradictory statements at p. 410]

Of Lyell and Darwin, Young (1973:386-387) says, "They are, relatively speaking, the purest of the scientists in the Victorian debate and as such are nearer to the positions of physicists, chemists and mathematicians."

These statements from Young's work are interesting not because his arguments in any way depend upon Darwin's motivational purity—they do not—but because they represent a revealing gesture towards an institutionalized orientation in the history of science.

Young's great contribution to Darwinism historiography has been to show how attention might be reoriented away from the individual evolutionist (Darwin) towards the "context" of evolutionary thought (Young, 1969). He has cogently argued for the necessity of setting Darwin's thought in the same context as evolutionary thinking in general in 19th-century Britain, and in the same context as social theory. Malthus's "influence" on Darwin is not so much demonstrated as assumed—on the basis of a persuasively argued "common context" occupied by both thinkers.

For all that Young's work has been important and convincing, he has not succeeded in shifting the balance of historical attention away from the individual and his motives. Thus, much of Young's argumentation is directed against writings like those of Sir Gavin de Beer where concern for Darwin's motivational purity, and for his freedom from taint by Malthus, is patently obvious:

> It is . . . clear that Darwin did not owe Malthus anything on the score of variation or natural selection, but only the realization that the high rate of mortality exacted by nature resulted in pressure. . . . The view that Darwin was led to the idea of natural selection by the social and economic conditions of Victorian England is devoid of foundation. [de Beer, 1963:100; quoted in Young, 1973:362-363]

And, more recently, de Beer reverses the question of intention to achieve the same effect:

> Malthus had not the slightest idea of natural selection and would have been horror-struck at the notion of evolution. What Darwin got from Malthus was something that Malthus knew nothing at all about, and about which he was not writing: how natural selection is enforced on plants and animals. [de Beer, 1970:33; quoted in Young, 1973:363]

Darwin is de-coupled from the suspect motives of Malthus since Darwin's "debt" resided in something of which Malthus "knew nothing."

An intriguing alternative interpretation is provided by Himmelfarb. She resolves the "oddity" of "the circumstance that a purely economic and political tract should have inspired a purely scientific theory" by a reinterpretation of Malthus:

> In fact, Malthus' theory was itself derived from natural history, and Malthus himself, although a political economist by profession, was a devotee of the natural sciences. [1968:160]

It was, Himmelfarb argues, the fact that Malthus's theory was *actually a piece of science* that solves the genealogical problem of impossible parentage—a relationship which is "one of the most curious and misunderstood in the history of ideas." Purity of ancestry is reestablished once it is understood that Malthus's influence was a properly scientific one. Indeed, Himmelfarb's argument goes further, absolving Malthus from Darwinian intentions, and asserting the purity of the *Malthusian* germ-plasm:

> This concept of a "struggle for existence" . . . represents the high point of Darwin's debt to Malthus. For the rest, however—the interpretation put upon it and the use he made of it—Malthus bears no responsibility. It was neither Malthus' intention to argue nor the effect of his argument that in these "struggles for existence" the strong triumphed and the weak succumbed. . . . It is likely that Malthus would have been much distressed had he lived to see what use Darwin made of him. [Himmelfarb, 1968:161, 163]

The most comprehensive recent defence of Darwin has been presented by the anthropologist Derek Freeman, who leaves ancestry, internal state, and germ-plasm all 'whiter-than-white.' Freeman's paper contends that Darwin 'had' the theory of natural selection before reading Malthus in 1838. Darwin's own acknowledgment of Malthusian influences was much too generous; all Malthus provided was a parturitional lubricant:

> That Darwin received a crucial stimulus from his reading of Malthus is widely known, but, as the historical evidence indicates, it was a maieutic stimulus, and not a pervasive ideological influence; . . . and further, as De Beer . . . has documented in his analysis of Darwin's research notebooks for 1837-1838, Darwin had already become aware of the existence of a natural process directly comparable to artificial selection some months before he read Malthus. [Freeman, 1974:212]

All that Darwin owed to Malthus was the catalytic stimulus of a *mathematical* formulation—no taint attaching to mathematics. This line is variously explored by Vorzimmer (1969), Herbert (1971), Hodge (1972), and, in spectacular scholarly detail, by Schweber (1977).[2] Regarding Darwin's internal state, Freeman (p. 213) reports that he was "well aware" of a "decisive break" with Lamarckian notions; that he published his pangenetic (Lamarckian) theory "in a state of frustrated bewilderment" (p. 214); that he did not really "believe that natural selection could ever produce 'absolute perfection' " (p. 219); and that his "true feelings" about Herbert Spencer were at best ambivalent (p. 219). Unlike the speculative and deductive Spencer, Darwin was, we are told, concerned only with factual evidence and justified inferences from the facts, as exemplified by Darwin's allegedly sober rejection of Lamarck, and Spencer's credulous acceptance thereof (pp. 217-218). The purity of the germ-plasm is argued from the contention that Darwin's theory "as he published it in 1859, was authentically scientific" (p. 215). Nothing in the *Origin* provides any basis for social Darwinism, whilst little in Spencer's work provides the basis for anything else (pp. 215, 218-221):

> It is of importance, therefore, that contemporary anthropologists should have an accurate understanding of those concepts of Charles Darwin on which the modern biological theory of evolution is founded, and that these concepts should not be confused with the obsolete Lamarckian evolutionism of Herbert Spencer. [p. 221]

Work by James Allen Rogers hinges upon analyses of "what Darwin actually wrote" in the *Origin* as a way of addressing the connection between Darwin and social Darwinism. The crucial move here is the intercalation of a process of "misunderstanding" between Darwin's *intentions* and the social meaning of evolutionary thought. Rogers's strategy is interesting not so much for its originality as for the promise it has already shown of becoming a new orthodoxy in Darwin-glossing (Rogers, 1972; cf. Gale, 1972:332; Young, 1971a:444). The key contention is that *"Darwin was not responsible for this* [social Darwinist] *interpretation of his theory"* (Rogers, 1972:271). He was "as much amused as annoyed at such gratuitous interpretations of his theory" (p. 269); and, while Rogers recognizes that certain of Darwin's private papers (not to mention later publications) *did* express convictions of the social relevance of evolutionary theory,

> What Darwin wrote in his letter concerning the applicability of the struggle for existence to the social evolution of man was of no consequence for the interpretation of the *Origin of Species.* These letters contained his private speculations and were not necessarily a logical extension of his work on the theory of natural selection. [p. 275]

However, Rogers (p. 265) insists, it is a mistake to ignore the "coincidental rather than instrumental" effect of Darwin's writings on social Darwinism. This "coincidental" effect arose from Darwin's *metaphorical* manner of expressing certain key concepts in the *Origin* (p. 268; cf. Gale, 1972:323-324; Young, 1971a; Peckham, 1959:40). It was not what Darwin *believed*, or even, strictly speaking, what he *said*, that formed his contribution to social Darwinism; it was, rather, "the manner in which he said it" in combination with the illegitimate ideological "set" of Victorian *society*:

> Darwin, as discoverer, had the usual difficulty of transforming an established vocabulary to describe a new vision of the process of evolution. By using metaphorical concepts from Malthus and Spencer, Darwin made it more difficult to dissociate his new discovery from older patterns of social thought. [Rogers, 1972:268]

We are invited to see Darwin's concepts of the struggle for existence and of natural selection as the consequences of a "failure" to express the (presumably distinct) scientific theory in absolutely neutral language. His failure was regrettable, but understandable, and Rogers repeats that Darwin was not "responsible for the later interpretations of his theory" (p. 268). Similarly, the debt to Malthus consisted in metaphor —a mere linguistic ornament which can be contrasted as required with what Darwin 'really believed,' what his theory was 'in reality,' even his original vision of "the process of evolution" (as both Rogers and Freeman malappropriately term it). What is interesting here is not so much the use of a suspect account of metaphor as mere ornament, as the use of that account to reveal purity beneath the ornament, both in the belief and intention of the author and in the theory embodied in his book. Unfortunately, no reconstructions of Darwin's beliefs or his written materials have yet been presented which reveal them bare of their linguistic "decorations!"

Rogers's approach accomplishes, with less panache and virtuosity, the same effect as an earlier analysis, to be found in Gillispie's *The Edge of Objectivity* (1960; cf. also his brief comment on Freeman, 1974:224). This analysis apparently concedes the case and acknowledges many kinds of pollution of Darwin and his work. Malthusianism was not just embodied in a text that Darwin happened to read "for amusement"; in Victorian England it "became transmuted into the moral foundation of liberal individualism" (Gillispie, 1960:311). It was part of the *Geist* of Darwin's *Zeit*. As Darwin read Malthus's *Essay*, these were the connotations that would have been called to his mind. The Malthusian model did indeed suggest to Darwin "the idea of natural selection":

> That certainly did happen, but the influence of political economy was more pervasive than that. Its ambiance provided Darwin, not just with a solution to his problem, but with a language expressive of a set of expectations about the way the world works. [Gillispie, 1974:224]

The way in which Darwin elected to express his scientific ideas in the language of struggle, conflict, and individualism makes it "inconceivable that it could have been written by any Frenchman or German or by an Englishman of any other generation" (Gillispie, 1974:224).

So far the exposition might profitably be read as sociological history of scientific knowledge. It turns out to be the prelude to purification by fiat: scientific ideas are treated as Platonic essences, by virtue of which their purity is guaranteed against the effects of any mere contingency; pollutions may stick to the surface but they cannot penetrate inside. Thus, the theory of evolution by natural selection is not "less scientific" for being expressed in the language of Victorian social and economic discourse: "classical political economy" is "the environment rather than the motivation of the theory." Scientific ideas cannot be contaminated by the idiom in which they happen to be expressed: the idiom is culturally contingent; science, including Darwin's, is timeless and above context. It is the benightedness of the non-scientist—the myopia of those who mistake a manner of speaking for the essence of eternal truth expressed in that contingent manner—which accounts for social Darwinism. Darwin could not "stop at every paragraph to explain himself":

> And we for our part may define social Darwinism as the re-exportation into social science of a language quite speciously fortified with the deterministic vigor of natural science—opinions converted into truths through having traversed science. [Gillispie, 1960:343]

It would be appropriate now to turn to the remarks on Darwin and Darwinism which might have provoked this extended affirmation of their purity. But there appears to be only one significant recent contribution which has cast doubt upon the master's internal state, that of the American anthropologist Marvin Harris. Harris's comments on Darwin's motivation were made in the context of a book on the rise of anthropological theory, in a section assessing the relationship between Darwin, Spencer, and social Darwinism (Harris, 1968:114-123). The *Origin* was "much more than a scientific treatise":

> It dramatized and legitimatized what many people from scientists to politicians had obscurely felt to be true without themselves being able to put it into words. [p. 117]

Contra Himmelfarb, who argues that Darwin was " 'without taint of ulterior ideology' ":

> One may readily admit that Darwin was without the "taint" but not that he was lacking in ulterior ideology. Having attributed the inspiration for his greatest idea to Malthus, he could not have been unaware of the larger implications of a "struggle for survival." [p. 117]

Darwin is argued to have suppressed his views on man and society in the *Origin* for expedient reasons (p. 118), those views being highly unsavoury: he was "a racial determinist" (p. 122); he "had a deep ideological commitment to *laissez-faire*" (p. 122); his principles were identical to Spencer's in that he applied "social-science concepts to biology" (p. 123). Harris apart, one may have to go to Spengler and Nietzsche for expressions of views which so clearly link Darwin's theory to "the atmosphere of the English factory" or his motives to those of factory owners (cf. Himmelfarb, 1968:418).

Thus, we can conclude this brief survey by observing that Darwin's defence is far better staffed and funded than its opposition. The only puzzle is why the trial has been conducted at all, and what hinges upon its outcome. What is to be made of the thematic importance of Darwin's internal state and the extended development of the language of "blame" and "responsibility?"

REFLECTIONS ON THE TRANSCRIPT

What might a professional anthropologist make of these phenomena were he to encounter them in a tribal setting? He would probably experiment with explanations in terms of myth-creation and sacralization. The scientific discipline of evolutionary biology had its font and origin in the person of Charles Darwin and in the text of 1859. Darwin is a sacred totem by virtue of his "foundership" of modern biology; science is sacred, so must Darwin and his Book be sacred; both must be protected from contamination by the profane. As the author of the *Origin* he must himself be pure; his thought must be unmingled with worldly pollutions and incapable of satisfactorily blending or combining with the suspect formulations of social Darwinism. Thus, "influences" from the "profane" Malthus can only be the spiritual emanations of mathematics and genuine science, or nonessential stimuli or manners of speech. And implications for social Darwinism can only be misunderstandings.

Of course, this is a wildly speculative suggestion. Nonetheless, it is interesting to note that references to "blame," "responsibility," and internal purity are less prominent where Darwin's work other than that recognized as his preeminent scientific achievement is the subject of study. This correlation of internal purity and high evaluation is just what the anthropological conjecture would suggest. There is, for example, comparatively little mention of internal states of belief, motive, or intention in Vorzimmer (1963), Olby (1966), or Geison (1969). These studies consider Darwin's accounts of blending inheritance, the inheritance of acquired characteristics, and pangenesis, using orthodox documentary methods to establish connections between texts. Geison's paper, for example, documents the similarities and differences between Darwin's "gemmules" and Spencer's "physiological units," and considers how far Spencer's work might have been used in Darwin's development of the pangenesis theory; it does this in a reasonably naturalistic way without insisting upon internal purity or even attempting to assess the extent of "pollution" (Geison, 1969:398-404). But Darwin's pangenesis theory, for whatever reason, is rarely treated as an authentic scientific achievement, or even as a paradigm of a properly scientific hypothesis. Similarly, Greene's account of Darwin as a social Darwinist (1977; cf. 1959:ch. 10; 1975:250) draws its evidence largely from unpublished writings and the *Descent of Man*, and not from an analysis of the *Origin*. The Darwin of 1859, the author of the *Origin*, remains set apart from the routine historical treatment which these contributions exemplify.[3]

It is interesting to note with what confidence the literature of purification describes Darwin's state of belief or intention. Evidently, nothing is more straightforward than the ascription of internal states to historical individuals. This probably reflects a more general and widespread set of historiographical preferences. Thus, intellectual biography, where the imputation of internal motivational structures is routine, is generally accepted as a securely based scholarly exercise; whereas the treatment of institutionalized patterns of thought, "movements," or "cosmologies" is frequently criticized as speculative and lacking in rigour. Since this is how things seem to stand, it will be worthwhile to cast a sceptical eye upon the whole process of identifying internal states. Historians rarely note that in philosophy and the social sciences the imputation of internal states of belief, motive, and intention is commonly treated as one of the most problematic and theory-laden of enterprises, and indeed is often condemned as illegitimate in principle.

The sociologist Wright Mills (1940) provides one of the most concise warnings against treating motives as anything but historically situated "vocabularies" used to "make-out," justify, and excuse behaviour (see also Coulter, 1977). At the other extreme, Rodney Needham (1972) devotes an entire book to discovering for anthropologists Wittgenstein's (1968) warnings about the perils of treating beliefs as internal states. In academic philosophy an extensive literature has developed the themes and insights of Wittgenstein's work, and demonstrating the inaccessibility of internal states now has practically the status of a specialty within the discipline (cf. Melden, 1961 for arguments of particular relevance in the present context). Academic philosophy might, perhaps, be thought too distant from the practical problems involved in understanding the thought of the individual, but even where all the resources of computer technology, statistical method, and psychological experimentation have been brought to bear, researchers have been daunted by the problems of measuring belief (Suppes, 1974). It is clear that radically opposed conceptions of what is sound and what speculative, what is straightforward and what problematic, can be sustained in different fields. Nonetheless, it is regrettable that descriptions of the psyche are so uncritically and routinely accredited by historians, especially since they tend to clash with what still surely must be the normal historical practice of concentrating on visible materials.

This is not to suggest that historians should abandon all interest in internal states. There are many contexts in which they would be entirely justified in ignoring a priori technical objections, and deploying the language of belief and motivation as part of a psychological *theory*. Courageous psychological theorizing, for all its risk and eccentricity, is certainly worth a try; Gruber and Barrett's work on Darwin provides a valuable and provocative instance (1974). The treatment of internal states by linguistic philosophers and other critics is not conclusive, nor, indeed, is it always technically satisfactory.

Nonetheless, the sober, orthodox philosophical treatment of the language of psychic states certainly seems of relevance here. As Mills (1940) says, vocabularies of belief and motivation are to be treated as shared public resources, intelligible by reference to the community possessing them, and what that community does with them. Thus, the imputation of motives by professional historians is to be understood by reference to *their* communal concerns and modes of practice. And, analogously, Darwin's self-imputation of motives can only be considered in relation to communal concerns and practices in the context wherein they were uttered.[4] Darwin's expressed motives do not provide a pipeline to his soul.

Let us expand the first point and concentrate upon the motives and intentions imputed by modern historians. Rather than taking motives as internal, antecedent causes of behaviour, we are to treat them as information which makes the behaviour meaningful and intelligible as a socially recognized kind of action (Melden, 1961:ch. 9). The ascription of motive evokes an orientation to behaviour, states what that behaviour is, makes the behaviour visible and intelligible as a category of some kind of action, and, thus, as a publicly significant "known" event.

What this amounts to in the case of Darwin studies is not a simple matter; nonetheless, we shall suggest what might be involved in the case of the motive ascriptions we have reviewed. The purity of Darwin's motivations is not now to be taken as an antecedent factor which helps us to understand his work as a naturalist. Rather, it is a way of making stipulations about that work. It makes it visible as properly scientific work, as the work of an ideally constituted producer of knowledge. In other words, the display of Darwin's purity of internal state is nothing but a way of making him out as the ideal-type of a modern scientist.

We can now see that anthropological speculation about purification-rites is broadly compatible with the sober doctrines of linguistic philosophy. Indeed, assessed in terms of these doctrines, it is *less* speculative than historical work relying upon the imputation of internal states. Thus encouraged, an anthropologist might be tempted to reformulate all the work on Darwin's purity along similar lines. The inessential character of Malthus's "influence" is not a report on the discernible connections of ideas, but a stipulation of the *appropriate* communal method of treatment of the ideas. The "essential truths" underlying Darwin's "mere" metaphors are not descriptions of, or even references to, what remains in the *Origin* when the language is "peeled away"; rather, they are means of locating Darwin's work in relation to *historians'* methods: they give the *Origin* status as proper science, and thus align it for study by the methods of the history of science proper, rather than as (say) "popular culture" or "ideology."

It may be objected that *any* historical study must centre upon internal states—even if not upon their essential purity. Despite difficulties, pitfalls, and temptations to fanciful moralizing, it might be argued that there is no alternative to such a focus. This is mistaken. There are alternatives; indeed they already exist in the literature on Darwinism. These approaches link readily with normal procedures of historical scholarship centred on documentation and the visible evidence of the written word. Let us give an example of one such approach. Traditional procedures frequently give a detailed indication of the relationship of an individual's writings to the writings of others, and make possible a reconstruction of a pattern of connections. (We have already noted this type of work in relation to Darwin's development of a genetical theory [e.g., Vorzimmer, 1963, 1970; Geison, 1969].) As the writings of more and more individuals are studied, the emerging pattern comes closer and closer to that of the overall development of the written culture. And we can begin to perceive general characteristics in such developments. For example, studies of the writings of individual authors have led to the identification by historians of the movement they have called "scientific naturalism"(Turner, 1974).

Historians' methods may dictate that the overall pattern is identified by the study of individuals' writings and biography, just as methods in cerebral anatomy dictate that the overall structure of the brain is identified through the consideration of numerous slices through its tissue. But it is not necessary to see the pattern of cultural change as an aggregation of individual contributions any more than it is necessary to see the structure of the brain as constituted by layers of tissue slices. Certainly, for many purposes, the part of the pattern of cultural change which happens to be associated with any particular individual may be quite unimportant. In seeking a general understanding of the rise of naturalism, for example, one can simply treat individual contributions as they fall into the pattern.

Individual writings can properly be expected to be connected in a pattern. No man is an island; no scientist, however cloistered at Trinity, Cambridge, or Down House, invents his own culture, his own language. We think and act on the basis of the resources our culture provides; and thereby we may add to these resources, and provide additional materials upon which the thought and activity of others may be constructed. There is no reason for treating Darwin, Newton, or indeed any scientist, as an exception to this. The use of cultural resources to construct scientific theories is an observation of what happens, not the casting of an aspersion.

Nonetheless, a culture need not be seen as an homogeneous whole. How differentiated it is at any time, how many "sub-cultures" exist within it, how "insulated" they may be from other sub-cultures and from the wider culture, and to what extent any individual's writings connect with this or that part—all these questions cannot be decided by a priori conceptions of how science *must be*, but only by "going and looking." Individualistic accounts of science, which relate the truth of science to its purity, anticipate the answers to empirical questions. What does it matter whether a scientist constructed his theories out of this or that cultural material, or in this or that psychological state? A theory may nonetheless be incorporated into recognizably "scientific" culture. And it may still be highly valued in terms of any particular standard of judgment.

In fact, there is every reason to suspect that the British intellectual scene of approximately 1859 was comparatively undifferentiated, with cultural resources being taken up from a common stock and put to a wide range of uses. Robert Young has convincingly documented this "common context" (1969, 1970, 1971a), and his observations are supported by the work of Greene (1977) and Gale (1972). Greene's account of Darwin's views on social evolution sets Darwin in a cultural context within which it was normal for "biological" and "sociological" materials to intermingle. But perhaps the most percipient observations come from the lesser-known work of Gale which suggests that "this was a period during which scientific, intellectual, cultural, social, and religious concerns and commitments were often indistinguishable" (1972:344), and that we ought to think, not so much of Malthus's "influence" on Darwin, as of Darwin's use of the materials which the undifferentiated culture provided for him (p. 343).

To the extent that Young, Greene, and Gale (and, for that matter, Gillispie) prove to be right about the context, we should stop making distinctions inappropriate to that context. As one consequence of this, Gale's work (p. 321) would cease to be (as he terms it) "a study in the extrascientific origins of scientific ideas." Another consequence might be that the "problem" of social Darwinism would disappear. If an examination of the written culture revealed no differentiated substructure which could appropriately carry the label, the alleged phenomenon would fall back into context as an aspect of the rise of naturalism. Certainly, it would be sad thus to lose a favourite topic of debate. But our boundary disputes are none of history's business.

NOTES

1. It may be as well to make clear what this review does *not* attempt to do:

(a) To notice the whole (or even a very large portion) of the secondary literature on Darwin, Darwinism, and social Darwinism produced in the past twenty years. Indeed, one of the most detailed studies (Manier, 1978) was not available to us at time of writing;
(b) to pretend to be written from a position of expertise in Darwin research itself;
(c) to attempt to *evaluate* the scholarly adequacy or factual accuracy of this literature.

2. Slightly off the beaten track of "Malthus-influence" studies, there has been the intriguing suggestion (attributed to Eiseley and Himmelfarb) that Darwin intentionally used his expression of "debt" to Malthus in order to disguise his "real debt" to other authors, and particularly to the naturalist Edward Blyth, from whom he is alleged to have 'pilfered' the principle of natural selection. A review of this line of argument is provided by Schwartz (1974), who claims that such innuendoes stem from "obvious malice toward" Darwin (p. 313), and who defends the purity of Darwin's citation-behaviour: "Malthus earned the credit he received from Darwin" (p. 306); Darwin did give Blyth "proper credit" (p. 314). Further developments in this area may have to await an accountant's audit.

3. We may amplify our contentions regarding the difficulty of ascribing motives by noting that the 1859 *Origin* is routinely accepted to be quite free from social exhortation. Nonetheless, this "fact" does not by any means close off the possibility of attributing "ideological" motives to Darwin. For example, the following possibility would seem to have as much to recommend it as other forms of "motive-ascription": Darwin was primarily motivated to advocate "social evolutionary" views; he calculated that doing so directly in 1859 would have had a counterproductive effect; he therefore couched his "ideological" vision in the "value-neutral" language of natural science; the *Origin* should thus be seen as a stratagem in an ideological "game" (cf. evidence on this in Greene, 1959:ch. 10). We do not *recommend* this account; we merely point it out as an option available to those who speculate on internal states; other options arise from other attributions of motive.

4. This is why there are immense problems in lifting motives out of context, as is routinely done with Darwin's in the literature of purification. Historians, with their professed aim of recreating past contexts, will see the force of this point.

REFERENCES

BOWLER, P. (1978) "Hugo de Vries and Thomas Hunt Morgan: the mutation theory and the spirit of Darwinism." Annals of Sci. 35:55-73.

BURROW, J. (1966) Evolution and Society: A Study in Victorian Social Theory. Cambridge: Cambridge Univ. Press.

COULTER, J. (1977) "Transparency of mind: the availability of subjective phenomena." Philosophy of the Social Sciences 7:321-350.

de BEER, G. (1970) "The evolution of Charles Darwin." New York Review of Books (December 17):31-35.

_____(1963) Charles Darwin: Evolution by Natural Selection. London: Nelson.

DURANT, J. (1977) "The meaning of evolution: post-Darwinian debates on the significance for man of the theory of evolution." Ph.D. dissertation, Cambridge University.

FREEMAN, D. (1974) "The evolutionary theories of Charles Darwin and Herbert Spencer." Current Anthropology 15:211-237.

GALE, G. (1972) "Darwin and the concept of a struggle for existence: a study in the extrascientific origins of scientific ideas." Isis 63:321-344.

GEISON, G. (1969) "Darwin and heredity: the evolution of his hypothesis of pangenesis." J. of the History of Medicine 24:375-411.

GILLISPIE, C. (1974) "Comment on Freeman," p. 224 in D. Freeman "The evolutionary theories of Charles Darwin and Herbert Spencer." Current Anthropology 15.

_____(1960) The Edge of Objectivity. Princeton: Princeton Univ. Press.

GREENE, J. (1977) "Darwin as a social evolutionist." J. of the History of Biology 10:1-27.

_____(1975) "Reflections on the progress of Darwin studies." J. of the History of Biology 8:243-273.

_____(1959) The Death of Adam. Ames: Iowa State Univ. Press.

GRUBER, H.E. and BARRETT, P.H. (1974) Darwin on Man: A Psychological Study of Scientific Creativity. New York: E.P. Dutton.

HARRIS, M. (1968) The Rise of Anthropological Theory. New York: Thomas Y. Crowell.

HERBERT, S. (1971) "Darwin, Malthus and selection." J. of the History of Biology 4:209-218.

HIMMELFARB, G. (1968) Darwin and the Darwinian Revolution. New York: W.W. Norton.

HODGE, J. (1972) "England," pp. 3-31 in T. Glick (ed.) The Comparative Reception of Darwinism. Austin: Univ. of Texas Press.

HOFSTADTER, R. (1944) Social Darwinism in American Thought. Philadelphia: Univ. of Pennsylvania Press.

LOEWENBERG, B. (1965) "Darwin and Darwin studies, 1959-1963." History of Sci. 4:15-54.
MANIER, E. (1978) The Young Darwin and His Cultural Circle. Dordrecht and Boston: Reidel.
MELDEN, A. (1961) Free Action. London: Routledge & Kegan Paul.
MILLS, C.W. (1940) "Situated actions and vocabularies of motive." Amer. Soc. Rev. 5:904-913.
NEEDHAM, R. (1972) Belief, Language, and Experience. Oxford: Basil Blackwell.
OLBY, R. (1966) Origins of Mendelism. New York: Schocken.
PECKHAM, M. [ed.] (1959) The 'Origin of Species' by Charles Darwin: A Variorum Text. Philadelphia: Univ. of Pennsylvania Press.
ROGERS, J. (1972) "Darwin and social Darwinism." J. of the History of Ideas 33:265-280.
RUSE, M. (1974) "The Darwin industry—a critical evaluation." Historyof Sci. 12:43-58.
SCHWARTZ, J. (1974) "Charles Darwin's debt to Malthus and Edward Blyth." J. of the History of Biology 7:301-318.
SCHWEBER, S. (1977) "The origin of the *Origin* revisited." J. of the History of Biology 10:229-316.
SMITH, R. (1972) "Alfred Russel Wallace: philosophy of nature and man." British J. for the History of Sci. 6:178-199.
SUPPES, P. (1974) "The measurement of belief." J. of the Royal Statistical Society, series B 36:160-191.
TURNER, F.M. (1974) Between Science and Religion: The Reaction to Scientific Naturalism in Late Victorian England. New Haven: Yale Univ. Press.
VORZIMMER, P. (1970) Charles Darwin: The Years of Controversy: The 'Origin of Species' and Its Critics 1859-1882. Philadelphia: Temple Univ. Press.
——— (1969) "Darwin, Malthus and the theory of natural selection." J. of the History of Ideas 30:527-542.
——— (1963) "Charles Darwin and blending inheritance." Isis 54:371-390.
WITTGENSTEIN, L. (1968) Philosophical Investigations. Oxford: Basil Blackwell.
YOUNG, R.M. (1973) "The historiographic and ideological contexts of the nineteenth-century debate on man's place in nature," pp. 344-438 in M. Teich and R.M. Young (eds.) Changing Perspectives in the History of Science. London: Heinemann.
——— (1971a) "Darwin's metaphor: does nature select?" Monist 55:442-503.
——— (1971b) "Evolutionary biology and ideology: then and now." Sci. Studies, 1:177-206.
——— (1970) "The impact of Darwin on conventional thought," pp. 13-35 in A. Symondson (ed.) The Victorian Crisis of Faith. London: S.P.C.K.
(1969) "Malthus and the evolutionists: the common context of biological and social theory." Past and Present 43:109-145.

6

THE RECEPTION OF A MATHEMATICAL THEORY: NON-EUCLIDEAN GEOMETRY IN ENGLAND, 1868-1883

Joan L. Richards

Looking back on the subject in 1897, Bertrand Russell saw 19th-century investigations into the foundations of geometry as forming a three-stage process. The earliest stage is marked by the "founders" of non-Euclidean geometry who worked in the early decades of the 19th century: Gauss, Lobatschewsky, and Bolyai. These men experimented with the possibilities of constructing geometries after replacing the parallel postulate with another; in essence their work involved testing the independence of the postulate. The mathematical community at large virtually ignored this work in the 1820s and early 1830s when it was being done.

The second stage, which has been characterized as the "metric stage," was introduced by the work of Bernhard Riemann and Hermann von Helmholtz in the 1860s. Their studies were motivated by an attempt to clarify the bases—logical and experiential—of our

AUTHOR'S NOTE: This chapter presents a brief overview of issues which are currently being researched for a doctoral dissertation to be completed in 1979. I wish to thank the following people who have been helpful with their criticisms and comments on drafts of this paper: Erwin Hiebert, Barbara Buck, Lorraine Daston, Jeremy Gray, Shirley Roe, Silvan Schweber, and Niel Wasserman. Of course any errors it might contain are my own.

spatial conception. Beginning with the general notion of a number manifold or "multiply extended magnitude," Riemann and Helmholtz explored the ways in which local properties could affect the structure of space. They found that Euclidean space could be formed by imposing a certain structure on the number manifold, but that non-Euclidean geometries could also be formed. Thus, they came to a realization of non-Euclidean geometries as had Gauss, Lobatschewsky, and Bolyai earlier in the century, but by a different route.

In the 1860s, at the same time that Helmholtz and Riemann were publishing their ideas, the synthetic approach to this field was reintroduced. Gauss's correspondence with Schumacher, in which he developed his synthetic non-Euclidean ideas, was first published between 1860 and 1863; the Hoüel translation of Lobatschewsky's work appeared in French in 1866. There was a confluence of the important strains of non-Euclidean thought in the late 1860s; a general awakening to earlier synthetic developments coincided with the introduction of newly developed metric geometry. For the first time mathematicians in general became aware of geometries which were different from that so neatly expounded by Euclid.[1]

In the ensuing decades of the century, mathematicians continued to explore non-Euclidean geometries, but in doing so, for the most part, they did not continue either within the synthetic or the metric traditions. Rather they focused on the possibility of introducing a metric into a projective space. That this could be done to generate Euclidean space from projective space had been first recognized by Arthur Cayley in his "Sixth Memoir on Quantics" (1859). In 1871 and 1873, Felix Klein demonstrated that Cayley's way of defining distance could be generalized to generate non-Euclidean as well as Euclidean spaces. Riemann's metric approach to space through number manifolds and local properties was replaced by this new approach; the third "projective" stage had begun.

Bertrand Russell was an Englishman, writing just at the end of the century he was describing. His perception of how geometry developed in the late 19th century might be challenged in detail, but is invaluable as an indication of contemporary impressions of what was happening in the field. It brings into sharp focus a tantalizing problem for the historian of this period. For Russell assigns the work of Riemann and those who developed his approach to geometry to a comparatively insignificant position as a transition leading into projective developments. From a modern standpoint differential geometry and the results inherent in Riemann's work are of great significance. The question of why the differential approach was not developed seems a complicated one and, particularly for those who look at the period through the lens of modern developments, it seems very difficult to understand.[2] The English had access to Riemann's work in translation within five years of its German publication; Helmholtz was publishing his ideas in English journals. And yet the English did not develop their results.

In this case, developmental patterns within the history of mathematics are difficult to account for unless they are placed in a context of the broader realm of intellectual history. There is good evidence that in this instance, the neglect of metric geometry in favour of projective geometry can be in some measure accounted for by reaction to certain implications of the form of non-Euclidean geometry which was being developed by Riemann and Helmholtz.

In order to understand the kinds of implications which were attendant on non-Euclidean geometry in the 1870s, when Riemann's and Helmholtz's ideas were introduced, it is first necessary to outline the position which geometry held within English philosophical traditions. When this position is clear, the implications which were inherent in non-Euclidean geometry will be easier to understand. A discussion of philosophical traditions will also shed light on the conflicts which were developing in the 1860s and 1870s over the status of scientific knowledge. It was in the midst of this controversy that the new geometrical ideas were introduced, and they had important implications for these discussions. Within this context, the reaction of English mathematicians as a group, to geometrical developments, their tendency to develop geometry within the projective framework rather than the differential one, makes a great deal of sense. In large part, it represents a conservative reaction against the new geometry, an attempt to maintain the status quo against the broad impact of differential geometry.

GEOMETRY AND THE PHILOSOPHICAL TRADITION

The majority of English mathematicians who were active in the 1860s and 1870s were educated at Cambridge. When they attended the University, mathematics was the only subject included in the curriculum which was not specifically classical. Its unique position within Cambridge education was largely justified by the position it was seen to hold within a larger philosophical tradition. In the education of most English mathematicians, mathematics was presented as a study with important and far-reaching implications.

Bertrand Russell's *Foundations of Geometry* (1897: 1) suggests a perspective from which to look at geometry in this English tradition:

> Geometry, throughout the 17th and 18th centuries, remained, in the war against empiricism, an impregnable fortress of the idealists. Those who held—as was generally held on the Continent—that certain knowledge, independent of experience, was possible about the real world, had only to point to Geometry: none but a madman, they said, would throw doubt on its validity, and none but a fool would deny its objective reference. The English Empiricists, in this matter, had, therefore, a somewhat difficult task; either they had to ignore the problem, or if, like Hume and Mill, they ventured on the assault, they were driven into the apparently paradoxical assertion that Geometry, at bottom, had no certainty of a different *kind* from that of Mechanics—only the perpetual presence of spatial impressions, they said, made our experience of the truth of the axioms so wide as to seem absolute certainty.

Geometry held a role in philosophy which was important enough that changes in the interpretation of geometrical knowledge could have strong ramifications. Philosophy of science was seen as closely tied up with the metaphysics of many other areas, so a change in the status of geometry could have a very far-reaching effect.

In this 19th-century tradition, explorations of the ways in which man had come to knowledge in scientific fields were seen as directly relevant to ways he might come to knowledge in other areas. In fields such as politics or morals, man's knowledge was as yet incomplete, and so unsure that it did not adequately serve as a model of human knowledge. But within the sciences, genuine knowledge had been gained, and the methods by which man gained knowledge in the sciences could serve to point out the methods by which he could hope to succeed in other areas. For example, in the "Introduction" to the *Philosophy of the Inductive Sciences,* William Whewell (1847: 3) said, "we have reason to trust that a just philosophy of the Sciences may throw light upon the nature and extent of our knowledge in every department of human speculation." And the opening sentence of John Herschel's review of Whewell's work makes it immediately clear that, though he might disagree with Whewell about many issues in the philosophy of science, he also saw a direct connection between the kinds of knowledge which can be gained in the sciences and in other areas.

> If the moral and intellectual relations of man have ever been justly regarded as transcending in importance all other subjects of human interest, the necessary dependence of his duties and responsibilities on his natural faculties must render it impossible to appreciate or define the one [moral] without entering into a close analysis of the other [intellectual]. [Herschel, 1841: 177]

William Whewell and John Herschel were two of the most influential English philosophers of their day. They did not agree with each other on many issues in the philosophy of science. In fact, Whewell is usually classified as an idealist, Herschel as an empiricist. In one very important way, however, they were agreed in their philosophies of science: it was clear to both of them that in the quest for knowledge, man was able to find truth. For Whewell, the idealist, man had true knowledge inherent in him. Inherent in man's mind was the capacity to recognize and work out the basic conceptions, the "Fundamental Ideas," which were the bases of the various physical phenomena. This capacity guaranteed that the knowledge thus gained would be absolutely true; once man had recognized a Fundamental Idea, there was no question as to its truth. It was the belief in such an innate capacity to grasp directly the principles lying behind perceptual reality which marked Whewell as an idealist (see Butts, 1968; Ducasse, 1951).

More surprising, however, was that Herschel, the empiricist, emphatically agreed that man could gain true knowledge in the sciences, knowledge of the reality lying behind the mere appearances. Although in his interpretation man gained knowledge through experience, Herschel made it abundantly clear that man's knowledge was not restricted to that of the changing forms of daily life. Our minds, he argued, have been so constituted that in interaction with our environment they lead us to the real and permanent truths which underlie our existence:

> For among the infinite analogies which may exist among natural things, it may very well be admitted that those only are designated, in the original constitution of our minds, to strike us with permanent force, to embody around them the greatest masses of thought and interest, to become elaborated into general propositions, and finally to work their way to universal reception, and attain to all the recognizable characters of truth, which are really dependent on the intimate nature of things as that nature is known to their Creator and which have relation to their essential qualities and conditions as impressed on them by Him. [Herschel, 1841: 182]

Herschel placed such stress on this point, in fact, that in his mind it almost completely overshadowed the differences between his position and that of Whewell, as both interpretations of science led to the same possibilities of truth:

> And, perhaps, with this explanation both parties ought to rest content, satisfied that, on either view of the subject, the mind of man is represented as in harmony with universal nature; that we are consequently capable of attaining to real knowledge; and that the design and intelligence which we trace through creation is no visionary conception, but a truth as certain as the existence of that creation itself. [Herschel, 1841: 182]

Thus, on the important issue, that there was an underlying order to the world which men can aspire to know, Herschel and Whewell agreed. The route which people took to reach this knowledge was at issue, but there was no doubt that it was attainable.

For both Herschel and Whewell, geometry served to support the contention that man could attain knowledge of the order underlying reality. Geometrical knowledge was real knowledge. Its axioms and theorems were exact descriptions of space; in realizing them man had caught a glimpse of the real order lying behind the changing facade of perceptual reality. Geometrical truths were necessary and universal; knowledge of them was so complete and definite that it extended beyond situations which had been experienced to encompass realms for which there was no empirical evidence. The space we live in was the space described in Euclid's geometry. This was one thing man knew absolutely.

This view of geometrical truth is interesting for two different reasons. In the first place, geometry was not viewed as an abstract logical system. Geometry, although logically developed, was interpreted as a study intimately tied to the world in which we live. Secondly, this view set geometrical truth off from truth which could be gained through mere deduction from a series of experiments. Simple experimentation could not justify extending one's knowledge past the experiments performed, the experiences known. But the conviction of the truth of geometry stretched far beyond the limits of every man's experience.

The far-reaching power of geometrical truth tied the basic study of geometry very closely to a study of the real world. Yet as presented in Euclid and studied in schools, it was essentially a logical system wherein theorems were developed deductively from a given group of axioms. The explanation for the way that logically developed geometrical truth could correspond to the real world lay in the interpretation of the axioms as the essential basis of the conception of space. For Whewell, this idea was inherent in each man, and was necessarily and universally true; for Herschel, it was learned through experience, but had the same necessary and universal scope it had for Whewell. Each of these men perceived geometry to be the logical development and exploration of a basic intellectual concept, the truth of which was absolute.

This view differs fundamentally from one which holds geometry to be merely a logical system constructed from an arbitrary set of axioms and definitions. For within the 19th-century interpretation, geometrical axioms were not arbitrarily chosen logical starting points for an investigation. They had a powerful ontological status which derived from their close connection to the basic concept of space. It was in this context, for example, that Whewell discussed the status of axioms and definitions in geometry. He devoted an entire chapter to this issue in the *Philosophy of the Inductive Sciences,* taking great pains to argue against those who claimed that the results of geometry were truisms implied by a set of arbitrary definitions.

> [N]o one has yet been able to construct a system of mathematical truths by the aid of definitions alone; . . . the definitions which we employ in mathematics are not arbitrary or hypothetical, but necessary definitions; . . . the real foundation of the truths of mathematics is the Idea of Space, which may be expressed (for purposes of demonstration) partly by definition and partly by axioms. [Whewell, 1847: 101]

Thus geometry was not an abstract logical system; rather it was the development of a basic spatial conception. The power of the study lay in the fact that the system thus developed corresponded exactly to external reality—in our spatial conception we had come to a real understanding of the underlying structure of the world.

By tying geometry to the concept of space, one explained not only how the logically developed system had such a clear experiential application, but also asserted that the truths gained through geometry were absolutely true. Herschel and Whewell needed to justify geometry's claims to this status before they could extrapolate from it to the possibility that other fields could gain this position. It was important that they intrinsically differentiated geometrical from physical knowledge and isolated the factors which distinguished geometrical truth from empirical truth.

The notion of conceivability stood at the heart of this distinction. In experimental physics, one could conceive of a situation which would run counter to physical theory; in geometry this was seen to be impossible. For example, though the sun rises daily, and we firmly expect it will tomorrow, it is possible that it will not. That we can conceive of that eventuality is part of what marks this prediction as contingent. It also sets it out as being essentially different from a geometrical prediction. For not only did one not find a triangle whose angle sum was greater than 180°, it was impossible to conceive of it. This distinguished geometrical from physical truth. Our whole experience takes place in Euclidean space—the impossibility of conceiving it otherwise marked this statement as necessary and universal truth (Whewell, 1847: 58-60).

In the middle of the 19th century, Whewell's and Herschel's philosophies of science represented the most widely accepted and comfortable interpretation which guaranteed man's capacity to reach truth. There were a number of other philosophies of science developed which were less optimistic about man's powers. These were the strong empirical philosophies, such as those of Hume or Mill: these were the philosophies which claimed that all of man's knowledge was gleaned from experience and did not include Herschel's extra assumption that we are guided to truth by the constitution of our minds. The strong empiricists were not as sanguine about man's ability to attain real knowledge of a truth behind appearances as were Whewell and Herschel.

Theirs was a disturbing message, because by limiting the truth which could be found in science, it called into question man's ability to find truth in any area. Many people were able to dismiss such claims, however, because of geometry. Strongly empirical philosophers were forced to assign to geometrical truth the same truth status as that of empirically gained knowledge. However, theirs was a difficult task. It was difficult to persuade people that they could conceive of a triangle with an angle sum different than 180° without producing one. And that was something which none of them could do.

Thus, in presenting a unified picture of human knowledge, many English philosophers depended heavily on geometry. For this was the area in which man's claim to real knowledge was the strongest. This was the area where, in the final analysis, the empiricist arguments were weakest because they could not produce a conceivable alternative to Euclidean space. And so, it was largely by analogy with geometry that absolute truths in other areas were known to exist. This was made explicit by Whewell, who defended the study of geometry as a part of liberal education at Cambridge with the following statement:

> The peculiar character of mathematical truth is that it is necessarily and inevitably true; and one of the most important lessons which we learn from our mathematical studies is a knowledge that there are such truths, and a familiarity with their form and character. [Whewell, 1845: 163]

THE INTRODUCTION OF NON-EUCLIDEAN GEOMETRY

In the 1860s and 1870s a strong challenge to this prevailing philosophy was mounted by a group of young scientists who have been characterized as scientific naturalists. Like Herschel and Whewell, the scientific naturalists believed that scientific knowledge was the model for the knowledge man could gain in all areas. Pushing this belief further, they felt that the scientific method was an adequate method—in fact the only legitimate method—for solving all of the problems of human existence. What set them apart from both Herschel and Whewell, and put them in a threatening position, was their strict delimitation of the kinds of truth to which such a scientific investigation could lead. The scientific naturalists did not claim to be able to arrive at knowledge of the reality which lay beyond or behind our sense perceptions. Theirs was a strict phenomenalism which maintained firmly that all man could know of the world around him was the information received through the senses. As for the transcendental realities, the scientific naturalists maintained that they were unknown and unknowable. Without Herschel's assumption that God guided man in what he gleaned from his experimentation, there was no way that the impressions gathered through the medium of the senses could be guaranteed to be true.

In the recent secondary literature there have been a number of widely disparate interpretations of the nature of scientific naturalism and this period in English history. Most of them focus on Darwin's *Origin of Species* and the issues raised around this work. For example, Young (1969, 1970) interprets the issues as part of a broad ideological movement dating from the 1790s and cutting across a number of disciplinary boundaries. Turner (1974), on the other hand, interprets scientific naturalism largely as a political movement stemming from the desire of scientists to gain more status and power within English society. Most directly relevant to the reception of non-Euclidean geometry, however, is the work of Ellegård (1958) who closely analyzed the discussions of Darwin's work which he found in the periodical press. Ellegård found that among the educated classes, the theme of idealism and empiricism was the unifying thread which tied together the many and varied issues. This theme is also found in another, smaller discussion which took place at this time—the discussion of geometry.

Thomas Henry Huxley and John Tyndall, both biologists, are perhaps the most remembered proponents of scientific naturalism;

much of its major thrust came from Darwin's *Origin of Species*. But another spokesman for the view that the methods of science encompassed the legitimate ways man could come to knowledge in all areas, that the kinds of knowledge to which man could aspire in any area were limited by those he could reach through science, was the mathematician William Kingdon Clifford. Although Clifford is now not as well known as Huxley, or other scientific naturalists, among his contemporaries he acted as a potent force. He was the youngest of the proponents of the new scientific naturalism; he was the most iconoclastic and uncontrolled in his attacks on the traditional values and institutions of his compatriots. His ideas were powerful, his presentation of them equally so, and they appeared in many of the most well-known periodicals of the day. Clifford was also integral to the introduction of metric ideas of non-Euclidean geometry into England. He is the man who translated Riemann's "Habilitationsvortrag" into English and had it published in *Nature* (1873). Much of his own creative work can be interpreted as an attempt to develop these ideas.

Clifford the polemicist and Clifford the mathematician have, for the most part, been treated separately and scantily by historians who are either investigating the religious, epistemological, and moral issues of the day, or are pursuing the mathematics of the period. Taken as a whole, however, in the linking of scientific naturalism and mathematics, Clifford emerges as a significant persona who provides insight into the way that geometry developed in England. For in his polemical writings, Clifford spelled out clearly the drastic implications, dimly felt by his contemporaries, inherent in accepting metric non-Euclidean geometry: implications which stretched far beyond mathematics and even science, into psychology, morals, and ethics.

The young Clifford graduated from Trinity College, Cambridge, in 1867, the year that Riemann's "Habilitationsvortrag" was published in Germany. He was an outstanding student, finishing as second wrangler and second Smith's prizeman. The education that he received there was a conservative one, strongly emphasizing the cultural convictions that there were absolutes governing the order of existence in all areas, be they physical, social, or spiritual. The wider philosophical justification for the inclusion of mathematics in the curriculum was as a direct support for such an interpretation (Whewell, 1845: 163).

Clifford, though a mathematician, chafed under this system with its absolutes. His first popular lecture, "On Some of the Conditions of Mental Development," (delivered at the Royal Institution, Friday, March 6, 1868), was a frontal attack on the view that one would really be educated by learning and developing implications of absolute truths. Using an evolutionary analogy, he suggested that plasticity and flexibility in the consideration of all ideas is necessary in order that a people and a culture might survive the ravages of time. Conventionalities, fixed rules of conduct, definite forms of knowledge and fixed notions of truth or good, he maintained, serve only to solidify a people and a culture. They are not absolute or constant. If rigidity in these areas sets in too strongly, he argued, a nation will find itself in a position where it can only be "improved away," replaced or destroyed by other, more vital peoples and cultures. Clifford concluded his lecture with the characteristically startling aphorism: "In the face of such a danger, *it is not right to be proper*" (1868: 328).[3]

Differential non-Euclidean geometry, as developed by Riemann and Helmholtz, provided the key by which Clifford was able to escape from the stultifying absolutist interpretation of geometry. For in the geometrical works of these Germans, Euclidean geometry lost its apodictic status, and was no longer accepted as possessing necessary and universal truth. Riemann and Helmholtz maintained an empirical view of geometrical truth. For them, Euclidean geometry was the study of the space of our experience, not the development of a spatial conception. In their studies of Euclidean geometry, they concluded that our treatment of space as Euclidean was an empirically based conclusion.

Riemann's "Habilitationsvortrag," entitled "On the Hypotheses which Lie at the Bases of Geometry," opened with a proposal to explore the assumptions which lie at the basis of our spatial conception and the connection between them and our experience. He found from his considerations of multiply-extended manifolds, which he held to be the most general concept of spatial extension, that our conception of space is only one of several which could be derived. He went on to explain the implications he drew from this conclusion:

> But hence flows as a necessary consequence that the propositions of geometry cannot be derived from general notions of magnitude, but that the properties which distinguish space . . . are only to be deduced from experience. [Riemann, 1867 (Clifford trans., 1873: 14-15)]

Continuing, he proposed to investigate systems of simple "matters of fact" which would determine our space as Euclidean within the general number manifold. He was very specific about the implications of his investigation for the truth status of geometry:

> These matters of fact are—like all matters of fact—not necessary, but only of empirical certainty; they are hypotheses. [Riemann, 1867 (Clifford trans., 1873: 15)]

In the conclusion of his paper he suggested that in the realm of the infinitely large or infinitesimally small, areas lying outside of our experience, our hypotheses as to the Euclidean nature of space might not hold true.

Helmholtz, whose discussion of this form of geometry was the first introduction of it given to most Englishmen, and who, as he developed his ideas, published them in English, was primarily a physiologist. He saw his development of non-Euclidean geometry as evidence for his physiological view that space was an empirically based conception, developed in each individual through childhood experience, and his papers were largely arguments to prove that point (see Richards, 1977). Within general overviews of the history of mathematics, he usually is given credit for recognizing the group of rigid body transformations as basic to the metric $ds = \sqrt{dx^2 + dy^2}$ which Riemann had taken as an elementary fact. But within the broader context of the introduction of non-Euclidean geometry, Helmholtz's work had further impact. For much of his effort in his English articles is directed at demonstrating the conceivability of non-Euclidean space.

The relevance of this argument lies again in the peculiar status of geometrical knowledge. As was discussed above, the development of geometry was seen as the working out of a conception of space basic to all human experience. The power of the spatial conception, the assurance that it represented the truth about underlying reality, lay largely in the impossibility of conceiving it any other way. Helmholtz argued against this viewpoint, not merely by showing non-Euclidean geometry to be consistent, but by arguing that it was conceivable. In "The Origin and Meaning of Geometric Axioms" (1870), published in *Mind* in 1876, he described for several pages what one would experience were one transported into a pseudo-spherical space. His conclusion to this discussion was:

> These remarks will suffice to show the way in which we can infer from the known laws of our sensible perceptions the series of sensible impressions which a spherical or pseudo-spherical world would give us, if it existed. In doing so we nowhere meet with inconsistency or impossibility any more than in the calculation of its metrical proportions. We can represent to ourselves the look of a pseudo-spherical world in all directions just as we can develop the conception of it. Therefore it cannot be allowed that the axioms of our geometry depend on the native form of our perceptive faculty, or are in any way connected with it. [Helmholtz, 1876: 304][4]

From a modern standpoint, Helmholtz's careful descriptions of life behind convex mirrors or through prism glasses can be interpreted as consistency models. From his standpoint, however, consistency was not the issue: he was interested in conceptualization. He was carefully constructing models to show that we can *conceive* of non-Euclidean spaces. For if this is granted, then Euclidean geometry is not necessarily and universally true. Thus one can argue that our knowledge of space as Euclidean is empirically based.

This interpretation of geometrical knowledge enabled Clifford boldly to deny the possibility of transcendental knowledge in geometry. There was no longer a difference between the knowledge reached in physics, and that gained in geometry. Clifford makes this explicit in *The Common Sense of the Exact Sciences* (1885: 43) where he opens his discussion of "Space" with the statement "Geometry is a physical science." The truth to which one could aspire through this science was the same experientially limited truth which could be gained in the context of physics.

The effect of the new geometry on Clifford's metaphysical speculations, the position he saw it could play in a broad world view, is evident in a lecture delivered in 1872, "On the Aims and Instruments of Scientific Thought" (delivered at the British Association for the Advancement of Science, August 18, 1872). Here Clifford clearly set forth his conviction, characteristic of generations of English thinkers —that man's knowledge in all areas in human endeavour must come through the application of scientific thought. Scientific thought, as defined by Clifford, however, can never extend beyond the limits of what we have experienced. Consequently, it can never allow us to attain a final knowledge of an absolute or transcendental realm which lies behind the world as we know it:

In making this claim, Clifford was very specific about the role of non-Euclidean geometry in his thought:

> I shall be told no doubt, that we do possess a great deal of knowledge of this kind [theoretically exact], in the form of geometry. . . . If this had been said to me in the last century, I should not have known what to reply. But it happens that about the beginning of the present century the foundations of geometry were criticized. . . . And the conclusion to which these investigations lead is that, although the assumptions which were very properly made by the ancient geometers are practically exact . . . for such finite things as we have to deal with, and such portions of space as we can reach; yet the truth of them for very much larger things, or very much smaller things, or parts of space which are at present beyond our reach, is a matter to be decided by experiment. [Clifford, 1872: 137-138]

Thus Clifford espoused a view which would confine the scientist to formulating descriptive laws, allowing no transcendental or absolute status for the laws so formulated.

Clifford extrapolated from this philosophy of science into broader social and metaphysical issues, using the same reasoning which had for so long buttressed the speculations of Whewell, Herschel, and others like them. In the decade of the 1870s he published a number of articles and lectures which carefully spelled out his views on subjects far outside the ken of formal mathematics. Their titles, "The Ethics of Belief," "On the Scientific Basis of Morals," "The Influence upon Morality of a Decline of Religious Belief," give an idea of the kinds of problems Clifford took up. In these works, he not only developed his own strongly agnostic ideas, but also criticized other scientists who strayed into similar speculation, but believed science and her method could be used in support of more traditional views. Such men as James C. Maxwell, William Stanley Jevons, Peter G. Tait, and Balfour Stewart were all criticized, often by name, for trying to justify God or Christianity through scientific analogy.[5] Universally, the opinions Clifford expressed in his popular works represented frontal attacks on the treasured values and ideas of mid-Victorian Britain; very often the expression of the ideas was unrestrained and strong. His arguments were well known and widely discussed; they were powerful and difficult to refute within their context.

The empirical implications Clifford found in Helmholtz and Riemann's work were not peculiar to himself, but were discussed independently of him by many of his contemporaries (see, for example, Helmholtz, 1870, 1872, 1876, 1878; Jevons, 1871; Tupper, 1872; Land, 1877). The broader implications an empirical view of geometry would entail were also widely recognized and were the subject of metaphysical and theological debate (see, for example, Stephen, 1874; Ward, 1874) Thus, at the same time that Darwin's *Origin of Species* was posing a threat to the argument from design, geometrical developments seemed to be challenging traditions in English natural theology by undermining geometry's longstanding position as the bulwark for belief in necessary and universal truth. It was clear that the way one interpreted non-Euclidean geometry could have a real significance which reached outside of mathematics.

THE PROJECTIVE SOLUTION

The mathematical community, of which Clifford was a young but respected member, certainly was not unaware of these implications, nor were they immune to their impact. The period following the recognition of the work of the German metric geometers saw a number of addresses by British mathematicians to their colleagues and wider audiences on the subject of geometry. Their themes varied from addressing the perennial question of whether the study of pure mathematics is justifiable for its own sake, to the ways in which geometrical ideas should be introduced in schools. One of the issues which occurred and recurred in these far-flung discussions is that of necessary and universal truth. The problem of how to interpret Riemann's and Helmholtz's mathematical results surfaced repeatedly in these discussions.

The solution to the question of the interpretation of non-Euclidean geometry which arose from the discussions of the 1870s and early 1880s was a conservative one. Non-Euclidean geometries were recognized as consistent mathematical theories but not as descriptions of possible spaces. The attempt was made to preserve Euclidean geometry in its special position as the expression of a spatial conception which possessed necessary and universal truth.

This attempt was virtually impossible within the context of differential geometry. For the very starting point of Riemann's and Helmholtz's investigations was the effort to construct a geometry from general notions of magnitude. The very starting point of the investigation was one which basically conflicted with the view of geometry as the working out of a spatial conception. Riemann was trying to examine the logical structure of geometry, something which was impossible within the conservative spatial interpretation.

The kind of opposition generated by the starting point of the differential investigators is clearly enunciated by Samuel Roberts in an address to the London Mathematical Society in 1882. Here he attempted to lay to rest the doubts which had been raised in the usually philosophically neutral ranks of the mathematicians by the work of the "metric" geometers.

> Let me direct your attention to the following important remarks by Riemann. After mentioning his purpose of constructing the notion of a multiply extended magnitude, he says . . . : "It will follow from this, that a multiply extended magnitude is capable of different measure-relations, and, consequently, that space is only a particular case of a triply extended magnitude. But hence it follows, as a necessary consequence, that the propositions of Geometry cannot be derived from general notions of magnitude, but that the properties which distinguish space from other conceivable triply extended magnitudes can only be deduced from experience."
>
> Now, since the word "magnitude" already savours of geometry, I remark that we may substitute the word "manifoldness," or simply

> "quantity"; and, if continuity is implied, we may say "continuum." I do not know what metaphysician or mathematician has essayed to derive the propositions of geometry from general notions of magnitude, that is to say, of "manifoldness" or "continuum." [Roberts, 1882:11]

Roberts continued his discussion by briefly listing a number of philosophical interpretations of the spatial concept. He dismissed Riemann's attempts at logical examination with the comment: "Philosophers are at issue with the distinguished analyst [Riemann] on this point [the interpretation of the spatial concept]" (Roberts, 1882: 11).

Criticism of the fundamental assumptions of the differential approach to geometry was one of the ways to nullify claims that within this study spaces had been created which could rival the monolithic Euclidean construct. If the starting point of differential geometry were not the basis on which Euclidean geometry was constructed, it had no real impact to say that non-Euclidean geometries could also be developed from these notions. Euclidean geometry was firmly maintained to have its origin in our spatial conception. As long as this was accepted, the possibility of logically developing other geometries was not a threat to its necessary and universal status.

It was difficult, however, to maintain this position towards differential geometry because of Helmholtz's popular essays. The German physiologist had made a startlingly successful attempt to show that one could conceive of non-Euclidean spaces by describing what a person would perceive if he were actually placed in such a situation. His careful descriptions of what life would be like in non-Euclidean space were not easy to dismiss. And yet, if they were accepted at face value, they would destroy geometry's claims at real knowledge by removing from it the mark which distinguished it as necessarily and universally true. Helmholtz would have shown one could not only develop non-Euclidean spaces, one could conceive of them. Thus, English mathematicians, anxious to nullify the metaphysical claims of differential geometers, were forced to deal with the models which Helmholtz had created.

One solution to this dilemma lay in finding a characteristic of Euclidean geometry other than its conceivability which would set it apart from other surfaces of constant curvature. This was the approach taken by Stanley Jevons, one of the first to respond to Helmholtz's work. Jevons emphasized that inhabitants of a non-Euclidean world would find that in the infinitely small their world followed the rules of Euclidean geometry. This property, that in the infinitely small any space of constant curvature approaches Euclidean space, sets Euclidean space off from any other. In Jevons' interpretation it served to mark

Euclidean geometry as the real geometry. Through it, he claimed, inhabitants of a non-Euclidean world would be able to recognize Euclidean space as true. They would conceive of space in the same way we do despite their anomalous experience. They would conceive of space as Euclidean and translate their peculiar experience into Euclidean spatial terms (Jevons, 1871:481).

Arthur Cayley, in his presidential address to the British Association for the Advancement of Science (1883), also tried to show how non-Euclidean geometries could be interpreted without jeopardizing the truth status of Euclidean geometry. He set off to demonstrate how "without any modification at all of our notion of space, we can arrive at a system of non-Euclidean (plane or two dimensional) geometry" (Cayley, 1883:9). To this end Cayley presented two models of non-Euclidean geometry. The first is the one Jevons discussed involving two dimensional beings inhabiting the surface of a sphere. Cayley's interpretation is somewhat different than Jevons's, but his conclusion is essentially the same. His sphere dwellers could come to know of Euclidean space; that this was impractical in their world would have no effect on its transcendental truth.

The second model Cayley used to show non-Euclidean geometries interpreted through Euclidean space was one of changes in length:

> [C]onsider an ordinary, indefinitely extended plane; and let us modify only the notion of distance. We measure distance, say, by a yard measure or a foot rule, anything which is short enough to make the fractions of it of no consequence . . . ; imagine, then the length of this rule constantly changing . . . , but under the condition that its actual length shall depend only on its situation on the plane and on its direction. . . . The distance along a given straight or curved line between any two points could then be measured in the ordinary manner with this rule, and would have a perfectly determinate value: it could be measured over and over again, and would always be the same; but of course it would be the distance, not in the ordinary acceptation of the term, but in quite a different acceptation. . . . And corresponding to the new notion of distance we should have a new, non-Euclidean system of plane geometry; all theorems involving the notion of distance would be altered. [Cayley, 1883:10]

This model of non-Euclidean space is less difficult to present as one which required no modification of our concept of space. It was easy to see intellectually how such a model would be constructed. And yet, Cayley did not make any attempt to describe how it would feel to live this way. He succeeded admirably in constructing a model which was powerful within the usual notion of space.

This model, which interpreted non-Euclidean geometries as dependent on definitions of distance, was one closely tied with projective geometry. And it was in this approach to geometry that mathematicians found a way to accept the mathematical reality of non-Euclidean geometries without having to deal with the issues which the differential geometers had implied. Projective geometry made it possible for mathematicians to reject the empirical message of the differential geometers and yet say with Roberts: "I do not call in question the mathematics of these new theories or underrate the brilliant power exhibited in their development" (Roberts, 1882:11).

Projective geometry, loosely defined as the study of those aspects of figures which remain invariant under radial projections had been developed in France and Germany through the first part of the 19th century. It has been argued (Nagel, 1939) that the development of projective geometry led ultimately to the view of geometry as an abstract formal system rather than as a study of space. But his perspective obscures the views of many people actually involved in working on projective geometry. Although its development may ultimately have led to a purely formal interpretation of geometry, during much of the 19th century projective geometry was interpreted as a study closely tied to space.

Projective geometry had originated in attempts to find real correlates for some of the ideas, such as negative and imaginary numbers, which were generating such powerful results in analysis (see further, Daston, forthcoming). These notions, which had no correlates in Euclidean geometry, found interpretations through projective geometry. Simply by focusing on different aspects of the spatial conception, by looking at those properties of figures which remained invariant under projection, projective geometers had found geometric analogues for analytic abstractions. Thus, projective geometry was a new approach to the spatial conception which significantly broadened its power and scope.

In the 1860s and 1870s, projective geometry was being introduced into English education by a number of English mathematicians who had been studying under German projective geometers. These men, most prominently T. Archer Hirst and Olaus Henrici, were involved in an active campaign to change geometrical teaching in England. Among other things, they wanted to introduce synthetic projective ("modern") geometry into the curriculum. The arguments for wanting to introduce projective geometry into the curriculum involved further variations on the themes of spatial conception which were being discussed in relation to differential geometry and non-Euclidean spaces. The rationale behind introducing projective geometry into the English curriculum was similar to the rationale which had originally motivated its original development in France. The interest in projective geometry lay largely in its close connection with the spatial conception which had generated Euclidean geometry.

In projective geometry, the pure mathematician could make contact with spatial reality even in his most abstract researches. This is the point Henrici made in his address to the mathematics and physics section of the British Association for the Advancement of Science in 1883:

> [T]o be thoroughly at home in the highest theories of pure algebra requires some of the genius of men like Cayley and Sylvester who have founded, and to a great extent built up, modern algebra. But even they constantly make use of geometry to assist them in their investigations. [Henrici, 1883:497]

Henrici's view is corroborated by Sylvester's simple statement: "whenever I went far enough into any mathematical question, I found I touched, at last, a geometrical bottom" (Sylvester, 1869:9). An important selling point for the study of projective geometry was that it was a way to extend our spatial conception to encompass vast areas of analytic thought. At heart, it simply involved a different approach to the basic spatial conception Euclid had originally explored.

Projective geometry's stretching of the results generated from the spatial concept was not a challenge to the validity of that concept. Whewell had made it clear that the axioms and definitions of geometry could not completely encompass the idea of space, that new ones were always possible. Roberts stressed the continual possibilities for development of the spatial conception in his article against the metaphysics of the differential geometers (Roberts, 1882:18). Rather than being a challenge to the necessary and universal status of our spatial concept, projective geometry served as a tribute to the power and scope of the fundamental idea of space.

Although its connections with the basic spatial concept were clearly asserted and recognized, the relationships which were explored within projective geometry were very different from those of Euclidean geometry. In particular, since length is not a property invariant under projection, there is no essential metric in projective geometry. But, in his "Sixth Memoir on Quantics" (1859), Cayley developed a method by which one could assign a number to any two points which would act in the system like Euclidean distance. In 1871 and 1873, Felix Klein modified Cayley's method for defining Euclidean distances, and was able to generate distance relations yielding non-Euclidean geometries within the projective approach.

This interpretation of non-Euclidean geometry through different definitions of distance in projective geometry represented the perfect solution to the problem facing English mathematicians. The belligerently empirical differential geometry of Riemann, Helmholtz, and Clifford could be ignored as their mathematical results were reinterpreted through the philosophically conservative projective geometry. Cayley's speech to the British Association for the Advancement of Science marked the culmination and clearest expression of this solution. In it he specifically indicated as significant the philosophical issues raised by the differential geometers. He wrote: "the notion which is really the fundamental one (and I cannot too strongly emphasize the assertion) underlying and pervading the whole of modern analysis and geometry, that of imaginary magnitude in analysis and of imaginary space . . . in geometry" (Cayley, 1883:8), and urged metaphysicians interested in mathematical issues to turn their attention to that problem.

Bertrand Russell's work, *The Foundations of Geometry,* was essentially an attempt to answer the question thus posed by Cayley in 1883. In the fourteen years which separated him from Cayley, Russell saw mathematicians to have turned their interest away from the differential geometry of Riemann, Helmholtz, and Clifford, and to have focused their geometrical interest in projective geometry. Securely rooted in the spatial conception, projective geometry had provided the English mathematicians with a respite from the furore raised by their empirical colleagues. It allowed the acceptance of the mathematical consistency of non-Euclidean geometries "without pomp and blow of trumpet as a signal victory for the empirical school" (Roberts, 1882:10).

NOTES

1. The negligible reaction to earlier works in non-Euclidean geometry and the burgeoning interest in this area in the 1860s was more carefully described in Bonola (1912).

2. Kline (1972:922-923) raised this issue only to conclude: "mathematicians can readily be carried away by their enthusiasms."

3. For an accurate view of the flavour of Clifford's work, it is well to consult the original essays rather than relying soley on the collection, *Lectures and Essays.* This is important because the editors admit that in reproducing the articles, "certain passages have been omitted which we believe Clifford himself would have willingly cancelled, if

he had known the impression they would make on many sincere and liberal-minded persons whose feelings he had no thought of offending" (Clifford, 1879:i, 70).

4. This article is a translation of an address first delivered in German in Heidelberg in 1870. This version is printed in Hermann vol Helmholtz, *Vorträge und Reden* (1884: ii, 3-31). It was translated into English and printed in *Mind* (1876:301-321).

5. Clifford criticized Maxwell and Jevons in his essay, "The First and Last Catastrophe." His essay, "The Unseen Universe," which is a review of Stewart's and Tait's book of the same name, is very strong in its attack on their Christian views.

REFERENCES

BONOLA, R. (1912) Non-Euclidean Geometry. La Salle, Ill.: Open Court.

BROWN, A. (1947) The Metaphysical Society: Victorian Minds in Crisis. New York: Columbia Univ. Press.

BUTTS, R.E. (1968) William Whewell's Theory of Scientific Method. Pittsburgh: Univ. of Pittsburgh Press.

CAYLEY, A. (1883) "Presidential address." Pp. 3-37 in Report of the British Association for the Advancement of Science.

______ (1859) "Sixth memoir upon quantics." Philosophical Transactions 149:61-91.

CLIFFORD, W.K. (1885) Common Sense of the Exact Sciences. (K. Pierson, ed.) London.

______ (1879) Lectures and Essays. Leslie Stephens and Frederick Pollack (eds.), London.

______ (1875a) "The philosophy of the pure sciences, Pt. II, The postualtes of the science of space." contemporary Rev. 25:360-376.

______ (1875b) "The first and last catastrophe." Fortnightly Rev. 17:465-484.

______ (1875c) "The unseen universe." Fortnightly Rev. 17:776-793.

______ (1872) "On the aims and instruments of scientific thought." Macmillan's Magazine 26:499-512.

______ (1868) "On some of the conditions of mental development." Notices of the Proceedings at the Meetings of the Members of the Royal Institution of Great Britain 5:311-328.

DASTON, L.J. (forthcoming) "The physicalist tradition in nineteenth century French geometry."

DUCASSE, C.J. (1951) "William Whewell's philosophy of scientific discovery." Philosophical Rev. 60:56-59, 213-234.

ELLEGARD, A. (1958) Darwin and the General Reader. Göteborg: Gothenburg Studies in English, Vol. 8.

FORSYTH, A.R. (1935) "Old Tripos days at Cambridge." Mathematical Gazette 19:162-179.

HELMHOLTZ, H. VON (1878) "The origin and meaning of geometrical axioms (II)." Mind 3:212-225.

______ (1876) "The origin and meaning of geometrical axioms." Mind 1:301-321.

______ (1872) "The axioms of geometry." Academy 3:52-53.

______ (1870) "The axioms of geometry." Academy 1:128-131.

HENRICI, O. (1883) "Opening address to mathamatica and physical section of British Association for the Advancement of Science." Nature 28:497-500.

[HERSCHEL, J.F.W.] (1841) "Whewell on the inductive sciences." Quarterly Rev. 68:177-238.

JEVONS, W.S. (1871) "Helmholtz on the axioms of geometry." Nature 4:481-482.

KLEIN, F. (1873) "Uber die sogenannte Nicht-Euklidische Geometrie, Zweifer Aufsatz." Mathematische Annalen 6:112-145.

——— (1871) "Uber die sogenannte Nicht-Euklidische Geometrie." Nachrichten von der Königlichun Gesellschaft der wissenschaften zu Göttingen 17:419-433.

KLINE, M. (1972) Mathematical Thought from Ancient to Modern Times. New York: Oxford Univ. Press.

LAND, J.P.N. (1877) "Kant's space and modern mathematics." Mind 2:38-46.

NAGEL, E. (1939) "The formation of modern conceptions of formal logic in the development of geometry." Osiris 7:142-224.

PEARSON, K. (1936) "Old Tripos days at Cambridge as seen from another viewpoint." Mathematical Gazette 20:27-36.

RICHARDS, J.L. (1977) "The evolution of empiricism: Hermann von Helmholtz and the foundations of geometry." British J. for the Philosophy of Sci. 28:235-253.

RIEMANN B. (1867) "Uber die Hypothesen, welche der Geometrie zu Grunde liegen." Abhandlungen der Königlichen Gesellschaft der Wissenschaften zu Göttingen 13:132-152. (English translation by Clifford [1873] "On the hypotheses which lie at the bases of geometry." Nature 8:14-17, 36-37.)

ROBERTS, S. (1882) "Remarks on mathematical terminology and the philosophic bearing of recent mathematical speculations concerning the realities of space." Proceedings of the London Mathematical Society 14:5-15.

ROTHBLATT, S. (1968) The Revolution of the Dons: Cambridge and Society in Victorian England. London: Faber & Faber.

RUSSELL, B. (1897) An Essay on the Foundations of Geometry. Cambridge.

STEPHEN, J.F. (1874) "Necessary truth." *Contemporary Rev.* 25:44-73.

SYLVESTER, J.J. (1869) "Address." Report of the British Association for the Advancement of Science.

TUPPER, J.L. (1872) "Professor Helmholtz and Professor Jevons." Nature 5:202-203.

TURNER, F.M. (1974) Between Science and Religion. New Haven: Yale Univ. Press.

[WARD, A.] (1874) "A reply on necessary truth." *Dublin Rev.* 2nd series 23:54-63.

WHEWELL, W. (1847) The Philosophy of the Inductive Sciences. London.

——— (1845) Of a Liberal Education in General. London.

WINSTANLEY, D.A. (1947) Later Victorian Cambridge. Cambridge: Cambridge Univ. Press.

YOUNG, R.M. (1970) "The impact of Darwin on conventional thought." Pp. 13-35 in Anthony Symondson (ed.) The Victorian Crisis of Faith. London: Camelot Press.

——— (1969) "Malthus and the evolutionists: the common context of biological and social theory." Past and Present 43:109-145.

7

PHYSICS AND PSYCHICS: SCIENCE, SYMBOLIC ACTION, AND SOCIAL CONTROL IN LATE VICTORIAN ENGLAND

Brian Wynne

I am afraid that this 'nature' is herself nothing but a first custom, just as custom is a 'second nature.' —Pascal, *Pensées*

The suggestion that moral or social concerns may be expressed in the idiom of natural laws no longer seems extraordinary, even though it may be treated as a threat by some. It is taken for granted in this chapter that ideas about nature even in their most institutionalized form—that is, in scientific thought—may come to reflect contextual factors in one way or another. However, this hard-gained vantage point only heralds the beginning of challenging problems confronting the historian and sociologist of science. The important questions concern the ways in which different sorts of social factors interact with other factors, including empirical, technical, and intellectual ones, to produce different forms of scientific knowledge.

One of the most important problems in this area concerns the multiple uses of scientific concepts, especially as between scientific and political or moral contexts. One argument holds that this "social use of scientific concepts" is simply a case of knowledge being drawn from science to underline a point, without any reciprocal influence upon the context of use within science. However, it is increasingly clear that this may not always, or even typically, be the case (Forman, 1971; Young, 1973). It is now argued that one context of use can never simply be assumed to be prior to or independent of another, even if it is the context of natural science, and that interaction between contexts of use should always be examined symmetrically with causal connections in both directions being sought (cf. Shapin, 1979). Much will depend upon the specific social nature of and relationship between those contexts.

One useful way of approaching an elucidation of the relationships between cultural context and natural knowledge is via the postulate that different social groups articulate and develop their concerns and social interrelations through forms of knowledge which contain tacit meanings embedded in them, and which are in a process of continual renegotiation, defence, and development. This is not an uncommon position in anthropology and sociology; it is stated particularly clearly by Mary Douglas in her *Implicit Meanings* (1975). She (1975:281) asserts the connection between society and natural knowledge in uncompromising and completely general terms:

> Apprehending a general pattern of what is right and necessary in social relations is the basis of society: this apprehension generates whatever *a priori* or set of necessary causes is going to be found in nature.

I would like to discuss, in an unashamedly exploratory way, the extent to which natural scientific knowledge can be justly and usefully regarded as a form of displaced, tacit moral discourse. I shall discuss one example—that of late Victorian physics. Although the exploration will thereby be narrowly focussed, I offer in compensation the analysis of a concrete case, from a well-documented historical situation.

THE SCIENTIFIC CONTEXT: LATE VICTORIAN PHYSICS

The physics of the last quarter of the 19th century has attracted the attention of a number of historians of science (Sviedrys, 1970; Swenson, 1972; Brush, 1967; Wilson, 1971; Goldberg, 1970), but as yet probably not in the degree which it deserves. During this period, as in so many others, British physics (or "natural philosophy") was dominated by Cambridge and recent emigrés from Cambridge to the provinces. At Cambridge during this period were the men at the forefront of late Victorian physics—Lord Rayleigh (the third Baron Strutt), J. J. Thomson, G. G. Stokes, Joseph Larmor, James Clerk

Maxwell. Close behind in eminence came Balfour Stewart, who went from Trinity to the Chair at Owens College, Manchester (where he taught Thomson and sent him to Cambridge); P. G. Tait, who went from Peterhouse to the Chair of Natural Philosophy at Edinburgh; William Barrett, who returned from Trinity, Cambridge to the Trinity, Dublin Chair of Natural Philosophy; and G. F. Fitzgerald, Fellow of Trinity College with Rayleigh and Thomson, and their colleague in the Cavendish Laboratory. Sir Oliver Lodge was a close friend of both Fitzgerald and Thomson and, although not a Cambridge don, worked with them at the Cavendish for short periods. W. M. Hicks was a Fellow of St. John's, friend of his co-Fellow Larmor, and later Professor of Natural Philosophy at Sheffield.

Although these physicists spanned a quarter century or more of scientific activity, and differed in their detailed interests and attitudes, nevertheless there were certain fundamental tenets which they held in common, against various competing trends and movements in scientific thought. It was they who bore the standard of British Victorian physics well into the 20th century and who continued steadfastly to articulate their paradigm in the face of eventually irresistible pressure for change (Goldberg, 1970).

One way of conveying something of the distinctive style and flavour of the work of this school of physicists is via a discussion of one of their most important concepts, that of the ether. This, of course, has been a widely employed concept throughout the history of natural philosophy. But it must be stressed that it possessed a quite distinctive significance and mode of use in the present context; that this was the result of an active transformation of their received conception of the ether on the part of the Cambridge physicists; and that this transformed conception is not readily intelligible as being "required by the state of experiment and observation." In this period, which followed Maxwell's development of the electromagnetic field equations, many conceptions of the character and status of the ether appeared plausible. Indeed Maxwell's own ambivalent attitude to this, and similar scientific concepts, neatly symbolizes the point (Kargon, 1969).[1]

Perhaps the most fundamental shifts from the received view were the unification of ether and ponderable matter, and the transformation from material theories of the ether to exactly the reverse, namely ethereal theories of matter. J. J. Thomson (1908) himself observed that this was the nature of the transformation in which he had taken a leading part between 1885 and 1900. Doran (1975) has documented Larmor's central role in this transformation to a thoroughgoing electromagnetic conception of nature rooted in and unified by ether.

Lodge (1889) expressed this view cogently in his *Modern Views of Electricity*, as did Hicks (1895). Michelson (1899:75) expressed the prevailing consensus thus:

> Suppose that an ether strain corresponds to an electric charge, and ether displacement to the electric current, these ether vortices to the atoms—if we continue these suppositions we arrive at what may be one of the grandest generalisations of modern science—of which we are tempted to say it ought to be true even if it is not—namely that all the phenomena of the physical universe are only different manifestations of the various modes of motion of one all-pervading substance—the ether.

A central characteristic of the late Victorian ether was its *supremacy* over matter and its nonmaterial nature. It was "matter of a higher order" with "a rank in the hierarchy of created things which places it above the materials we can see and touch" (Fleming, 1902:191). According to Stokes (1883:17):

> We must beware of applying to the mysterious ether the gross notions which we get from the study of ponderable matter. The ether is a substance, if substance it may be called, respecting the very existence of which our senses give us no direct information; it is only through the intellect . . . that we become convinced that there is such a thing.

Larmor (1900:vi) was even more categorical when he explained that

> Matter may be and likely is a structure in the ether, but certainly ether is not a structure made of matter. This introduction of a suprasensual ethereal medium, which is not the same as matter, may, of course, be described as leaving reality behind us; and so in fact may every result of thought be described which is more than a record or comparison of sensations.

Matter, electricity, indeed all physical phenomena were to be made intelligible as properties of a "suprasensual" ether. Thus, the ether took on a transcendent, *unifying* role within science. It was simplicity lying behind diversity, coherence behind disorder. It established continuity and connection between disparate particular events. Larmor (1908:43) put it as follows:

> Our conviction of an orderly connexion between things constitutes the conception of a *cosmos*. We have placed the foundation of this in the existence of a uniform medium, the ether, the physical groundwork of interstellar space. . . . The only ground for postulating the presence of this medium is the extreme simplicity and uniformity of the constitution which suffices for its functions. Needless to say, there remain many unresolved features, some still obscure, but hardly contradictory. But should it ever prove to be necessary to assign to the aether as complex a

> structure as matter is known to possess, then it might as well be abolished from our scheme of thought altogether. We would then fall back on simple phenomenalism, proximate relations would be traced, but we need not any longer oppress our thoughts by any regard for a common setting for them; the various branches of physical science would cultivate with empirical success independent modes of explanation of their own, checked only by mutual conservation of the available energy, while the springs of their orderly connexion would be out of reach.

And Lodge (1913:15, 17) considered the aether to be:

> the universal connecting medium which binds the universe together, and makes it a coherent whole instead of a chaotic collection of independent isolated fragments.
>
> The ether of space is at least the great engine of Continuity. It may be much more, for without it there could hardly be a material universe at all.

The physicists were willing explicitly to acknowledge their *faith* in the all-unifying medium. Here they were reacting against positivistic and naturalistic competitors who sought to do away entirely with entities whose existence could not be empirically observed. Fitzgerald (1896:441) rebuked one such competitor thus:

> Professor Ostwald ignores such theories as that of vortex atoms, which postulate only a continuous liquid in motion; but, it may be, this is omitted because it is merely a way of explaining the atoms. He also ignores metaphysical questions, such as whether motion be not only the objective aspect of thought, and also whether an intuitively necessary explanation of the laws as distinct from the origin and consequent arrangement of phenomena is not postulated by the fact that the Universe must be intelligible. Consequently, his attempt to deal with nature in a purely inductive spirit is unphilosophical as well as unscientific. The view of science which he puts forward—a sort of well arranged catalogue of facts without any hypotheses—is worthy of a German who plods by habit and instinct. A Briton wants emotion—something to raise enthusiasm, something with a human interest. He is not content with dry catalogues, he must have a *theory* of gravitation, a *hypothesis* of natural selection.

Thus did Fitzgerald unashamedly defend a "metaphysical" science, basing itself in "necessary" a priori principles of reality such as the all-unifying ether as much as in empirical observables. Indeed, antipathy to empiricism and positivism extended right down to matters of detailed scientific method. The "Cambridge" style vigorously criticized overemphasis upon precision and decried the tendency in other approaches in physics to deny existence to anything which did not yield to highly accurate measurement. Nor was this a later ad hoc defence of ether against scientific progress, for even Rayleigh, perhaps the most utilitarian of this group, was emphasizing this point as early as 1884, against the naturalistic complacency of the previous decades which assumed that "all the really great discoveries had been made and that nothing remained but to carry them to the next decimal place" (d'Albe, 1923: 256). Lodge (1913: 9) as usual was most fluent in his encapsulation of the point:

> The simple laws on which we used to be working were thus simple and discoverable because the full complexity of existence was tempered to our ken by the roughness of our means of observation. . . . Kepler's laws are not accurately true, and if he had had before him all the data now available he would never have discovered them.

The Cambridge group championed, instead of precision, the *imagination* as paramount in going beyond the "cowardly security" of empirical sense-data. And they urged cultivation of "the neglected borderlands between the branches of knowledge" (Rayleigh, 1884) which naturalism was fragmenting, with its denial of any common setting or *plenum* in nature for developing specialist disciplines. Probably the most fruitful research programme of the era, Thomson's conduction of electricity through gases of greater and greater rarefaction (which led inter alia to discovery of the electron) was the outcome of such an approach, presaged by Crookes' (1879) famous investigation of "matter in a fourth state," or "radiant matter" where:

> We have actually touched the borderland where matter and force seem to merge into one another, the shadowy realm between known and unknown which for me has always had peculiar temptations. I venture to think that the greatest scientific problems of the future will find their solution in this borderland and even beyond; here it seems to me, lie ultimate realities, subtle, far reaching, wonderful.

These ultimate realities in the shadowy unknown would have been regarded by scientific naturalists as so much metaphysical baggage. But they were central to the Cambridge approach, which indeed increasingly, if paradoxically, felt the need for a scientific demonstration of their presence. This approach insisted upon the ethereal constitution of matter and its continuity with radiation, ontological realism, and an essentially transcendent continuity in nature which could be mirrored in a systematically connected natural philosophy. And all these themes were, of course, intimately bound to each other and mutually supporting. It remains to be asked why late Victorian physics should have manifested such a particular and distinctive pattern of concepts and princples. So far these concepts and principles have been discussed only in the context of use provided by the esoteric scientific forum of the time. When one examines the social context more generally, however, it is immediately apparent that the same concepts and principles enjoyed a useful life in other areas. And this raises the possibility that by broadening the scope of enquiry some progress towards answering this intriguing question might be made.

THE SOCIAL CONTEXT: LATE VICTORIAN CAMBRIDGE

The history of the 19th century in Britain is dominated by the episode of industrialization and its effects. By the end of the century the lion's share of political power and influence had passed to the main agents and beneficiaries of industrialization, the bourgeois middle class, at the expense of the traditional landed interest. Even the last bastions of traditional control—the services, the Church, and the universities—were exposed to the "vulgar," meritocratic invasions of liberalism, the political manifestation of industrialism (Haines, 1969; Rothblatt, 1968). An earthy, antimetaphysical materialism, and a fundamental individualism, found proponents even within the confines of Oxford and Cambridge.

One way in which this trend was manifested was, needless to say, in new conceptions of science and scientific education. There was a growing stress upon the vital role which a properly organized science could play in an industrial state, and demands for the incorporation of science into a thoroughgoing utilitarian frame of reference (MacLeod, 1972). Such demands were pressed increasingly by an emerging group of practising scientists, who advocated an explicit policy of scientific professionalization, and a consequent radical reorganization of the old Universities. For E.R. Lankester (1910), a typical "professionalizer," Oxford and Cambridge purveyed "an empty sham of ancient culture rather than. . .the real and inspiring scientific culture of the modern renaissance"(1910:7). They were nothing more than "a couple of huge boarding schools, which are shut for six months of the year" (p. 8), and it was "monstrous" and "disastrous" that they should so dominate the country's learning and values with a profoundly anti-utilitarian, anti-industrial ethos of knowledge.

Not only did the professionalizers, in the name of industrial and scientific progress, demand such material changes in the institutions of learning as to threaten the power base of the traditional upper classes, but they also posed a corresponding threat in the realm of ideology. Part of the social programme of professionalization under the utilitarian ethos required that a public image of neutrality and objectivity be cultivated. This in turn required that any traces of metaphysics be expunged from science, at least in its public image. The social correlates of this implied a power struggle, initially between the established Church and the professional vision of science, which imperialized no matter what realms of experience with its secular naturalist ethos.

Scientific naturalism was the ideology of the advocates of the freedom and domination of professionalized science. It was also, thereby, the cosmological backcloth of industrialization. The social and moral connotations of naturalism were that, since (mechanist-materialist) science could comprehend all there was to experience, then society must be persuaded to look towards rational, scientific, and secular ideas to solve its ultimate problems and interpret experience, rather than towards Christian or otherwise metaphysical or "anti-progressive" modes of thought (Turner, 1974b). Scientific men and institutions must replace their ecclesiastical counterparts as the leaders and educators of culture and political virtue. The professionalizers talked of "The Scientific Basis of Morals" (Clifford, 1875) and referred to "the scientific knowledge of life as the one sure guide and determining factor of civilisation" (Lankester, 1910:13). Destiny was to be placed firmly in human, scientific hands, not in divine ones (Turner, 1974a; Brown, 1973). And, whereas the divine hand had previously been ideological maiden to the landed aristocratic master, the newly raised hand of utilitarian and professionalized scientific materialism was maiden to the industrial bourgeoisie.

In opposition to this threat from the aggressively ascendant industrial middle class, there arose a world-view which attempted to destroy or at least mitigate that threat. A sense of general social crisis was expressed which was attributed to the debilitating consequences of industrialism and its materialist credo (Rothblatt, 1968; Maurice, 1882:674; 1892:459). And there was a stress upon metaphysical unity to give coherence to a moral universe and a society which was seen as falling apart under the atomistic nihilism of the scientific naturalists. The professionalizers and their cosmology were regarded as the prologue to utter social fragmentation and chaos. With their concern to decouple scientific knowledge from metaphysical entities, to leave each scientific discipline free to wander its own, increasingly specialized and fragmented, utilitarian pathway, and to deny any transcendent reality suffusing the vulgar material world, the professionalizers were advancing a model of the social role of knowledge which was reprehensible. It abandoned causes and meanings for mere descriptive laws and formulae. It claimed no wider role than technical usefulness, or, when it did claim a wider "moral" role, it was on the basis of comprehensive social planning rooted in middle-class materialism.

That Cambridge dons should be foremost in formulating this reaction to naturalism is hardly surprising given Oxbridge's central social position in the traditional pantheon, and given that its own institutions had now fallen under direct threat from the aggressive materialist tide. Rothblatt (1968) has described the "Revolution of the Dons" of the Cambridge colleges against the encroachments of the vulgar middle-class industrialists who had gained a power-base within the University itself. (The Cavendish physics laboratory had originally been set up under the University's auspices as a utilitarian, professionalized scientific training centre to counter the colleges' dominating ethos of culturally and spiritually inclined natural knowledge [Shaw, 1926; Sviedrys, 1970; "Cavendish," 1910]. However, the colleges fought back successfully and managed to "infiltrate" the Cavendish with their own champion, J.J. Thomson, who was perhaps the leading figure of late Victorian physics.)

The Cambridge intellectuals, Maitland, Seeley, Sidgwick, Maurice, and others (including their Oxford counterpart, Matthew Arnold) warned of the need for a new, unifying intellectual universe to underpin a revamped moral and political universe of unity and harmony. As Rothblatt put it (1968:119), they were out to find "a new sovereign authority which would end anarchy, discredit each man doing as he likes, put down the working classes and heal the social divisions." Believing that England was disintegrating under the vacuum of instrumental amorality embodied in the new scientific professionalism and industrial materialism, these intellectuals,

> The England of the aristocracy and the Church behind them,... examined every available intellectual theory in the hope of finding a guide for their actions. Every new or old idea that came to hand, scientific, historical, theological, logical, or aesthetic, was anxiously inspected for its possible prophetic content. [Rothblatt, 1968:244]

Above all they searched for a scheme which evoked a sense of organic unity and harmony (Maurice, 1882, 1886; Backstrom, 1974). Against the social managerial hubris of scientific naturalism, they emphasized the absolute importance of organic *system*, of factors which would heal the wounds of atomism and which lay beyond the manipulatory powers of fragmented and reductionist bodies of knowledge. Against the professionalizers' interest only in narrow instrumental capability, they stressed the need for unified conceptions of nature, to the extent of stressing the unseen, "spiritual" aspects of nature and experience. This was entirely consistent with a wider moral role for natural knowledge. Indeed, they explicitly advocated (e.g., Maurice, 1882; Seeley, 1882) the equation of science and religion in a new form of "higher law" administered by a scientific clerisy. Maurice (1882: 674) referred to the need that "for the good order of society and for the sake of avoiding possible perils hereafter, certain forms of expression must be kept up which seem to imply that the infinite can be brought within the range of our cognisance," and he explicitly included "the man of science, just so far as he testifies to right, order, truth, in the government of men and the operation of nature."

Clearly, the concepts of the organic unity of knowledge, metaphysical realism, and the unseen world articulated in general opposition to the ideology of naturalism are strikingly reminiscent of the themes and principles articulated by the late Victorian physicists in esoteric scientific contexts. The hypothesis of systematic, significant relationship must be very plausible. But what real evidence exists of the relationship,and what kind of relationship was it? These are difficult questions, but one thing they suggest is the need to examine how the physicists interacted with other aspects of Cambridge culture. This immediately leads to some striking support for the initial hypothesis.

One venture in which many Cambridge intellectuals were involved, indeed which was dominated by them, reflects many of the important undercurrents in the cultivation of natural knowledge at this time, and links more or less directly with physics. This was the field of psychical research. Part of the incipient disintegration which the upper classes felt (or felt it necessary to articulate) was fuelled by the "underground explosion" (Howe, 1972:61) of occultism and antirationalism which occurred from about 1870 onwards. It is no coincidence that this anarchism in the realm of knowledge was strongly linked to political anarchism. Knowledge appeared to be so bound up with society that to break down the intellectual order seemed to be identical to breaking the social order.

One such current of decentralist antirationalism was spiritualism, which flourished among the "lower orders" of society as a mélange of grass-roots politics, religion, and alternative science (Podmore, 1902; Crookes, 1874, 1926; Nelson, 1969; Hill, 1932; Metcalf, n.d.). At times patronizing towards this grass-roots celebration of the supernatural, and at other times more aggressively antagonistic,[2] the Society for Psychical Research (SPR) was formed in 1882 by Henry

Sidgwick, Edmund Gurney, Frederick Myers, and William Barrett the physicist (all fellows of Trinity College, Cambridge) to investigate the phenomena of spiritualism and related matters (Gauld, 1968; Nicol, 1972). It championed "the unseen world" explicitly as a weapon against the limitations of the materialist cosmology, and used an elaborate scientific approach to establish the scientific reality of events and agencies beyond the ken of that cosmology and its adherents. But it also sought "scientifically" to control, from its own upper-class social base, all excursions into the unseen world beyond current boundaries of ordered experience, so as to discredit the "morally degenerate" interpretations being derived from such excursions (Wynne, 1977).

The activities and intellectual approach of the SPR incorporated science as part of the weaponry of social negotiation of legitimate meanings and social relations. Ironically, psychical research turned out to be a naturalization of the supernatural—a form of *scientific* supernaturalism which attempted to trump the naturalism of the professionalizers with a more comprehensive, "transcendent" naturalism. It wanted scientific authority for its own metaphysical cravings and social ambitions. As Henry Sidgwick himiself put it, in thoroughly utilitarian terms (James,1970:7):

> What we aimed at from a social point of view was a complete revision of human relations, political, moral and economic in the light of the science directed by comprehensive and impartial sympathy and an unsparing reform of whatever, in the judgment of science, was pronounced to be not conducive to the general happiness. . . . [The] time had come for the scientific treatment of political and moral problems.

The difference from the utilitarianism of the industrialists and professionalizers was only that the social base of the reform, and the specific content and form of its ideological medium, science, was that of the upper classes. They too, under threat, were forced to abandon the authority of pure tradition for the authority of alleged empirical demonstration and natural law. No matter that the science concerned contained no meaningful social directives, except a vague concept of universal unity and "spiritual" direction of matter; it was the authority to socially interpret reality for society at large which was crucial.

It is worth outlining the intimate social connections between the upper-class Cambridge intellectuals, the leading members of the SPR, and the physicists who constituted the orthodoxy of the late Victorian period, to illustrate how difficult it would be for those physicists to be unaware of or uninterested in the wider conflicts and concerns of their social context.

William Barrett, physicist and founder member of the SPR with his fellow Trinity dons Sidgwick, Gurney, and Myers, was also a friend of Balfour Stewart, who was a Fellow of Trinity and later Professor of Physics at Owens College, Manchester. Stewart and P.G. Tait, Fellow of Peterhouse, and later Professor of Physics at Edinburgh University, wrote the anonymous book, *The Unseen Universe*, in 1874, and followed it up with *Paradoxical Philosophy* (1878). F.D. Maurice was Master of Trinity, Sidgwick's tutor and friend.[3] Sidgwick's wife was the sister of A.J. Balfour, Tory Prime Minister, ex-Trinity undergraduate, and member of the SPR. Balfour was made President of the B.A.A.S. in 1904 and in his Presidential Address made an eloquent philosophical plea on behalf of the ether. Sidgwick, Balfour, and Mrs. Sidgwick were all Presidents of the SPR in its first thirty years. Lord Rayleigh, the eminent physicist and inheritor of huge landed estates, married another sister of Balfour; was an undergraduate at Trinity with Balfour; became Master of Trinity College (after Sidgwick) and Cavendish Professor of Physics at Cambridge; and was also President of the SPR Sidgwick. After Mrs. Sidgwick acted as a research assistant for Rayleigh in the Cavendish laboratory. J.J. Thomson came to Cambridge from Owens College where he had been taught by Balfour Stewart, became a Fellow of Trinity, succeeded Rayleigh as Cavendish Professor of Physics, and also as Master of Trinity College. He became a close friend of Rayleigh and tutored his son, the 4th Baron Strutt. Thomson, although never President of the SPR, was a lifelong member of its Council. Significantly enough, he chose to devote two chapters of his autobiography (1936) to psychical and related research, defending it against its "professionalist" detractors. A.J. Balfour was an accomplished philosopher and cultivated an elite social-intellectual circle which included Sir Oliver Lodge (Jolly, 1974:115). Lodge was an ardent psychical researcher and an eminent scientist, being F.R.S. and Professor of Physics at Liverpool University. He was a friend of Sidgwick, Myers, Stewart, Barrett, and Gurney, and a close friend of J.J. Thomson, Balfour, and Fitzgerald. Another leading physicist-mathematician of this period was Joseph Larmor, Fellow of St. John's College, Cambridge, whose affairs he administered together with the biologist William Bateson. Larmor was a friend and close colleague of J.J. Thomson and Lodge.

It is clear therefore that the dominating elite of British physics at this time was actively involved in psychical research, which was itself so embroiled in controversy that they must have been aware of the wider implications of such involvement. It was not self-evidently a "natural" move to make. Furthermore, the physicists were intimately connected, socially and intellectually, with the elite of conservative politics and of moral and political philosophy. The two contexts, of social and political debate, and of scientific research, have been pulled closer together and rendered less clearly distinguishable. But still more than this can be established. Many of the writings of the physicists themselves were not addressed purely to esoteric audiences, and some involved obvious political statements. Moreover, in such writings the concept of the ether itself was employed in legitimating a particular vision of the social order.

In *The Unseen Universe* (1874), Stewart and Tait laid out, with all the authority of physics, a rigorous scientific model of the nonmaterial world and its hierarchical connections with the different levels of the material world. It was a deliberate attempt to justify immortality and the superiority of Spirit. It cut into the defensive parallelism between God and nature implied in natural theology, by attempting to demonstrate that Divine providence operated through new and higher natural laws. In this work the ether appears as "not merely a bridge between one portion of the visible universe and another," but "a bridge between one order of things and another" (1874:158).

Stewart, an SPR member, also wrote a paper which appeared as an appendix to Barrett's book *On The Threshold of the Unseen* (1895) in which he sought a "higher law" than the materialists could see in nature. It was a cardinal tenet that this "higher law"—of manifest moral significance—would make itself known in as experimental science, which was why the SPR's activities were "of unusual importance." Spiritual realities would be incorporated in a higher natural law, and thus be demonstrable through science. This was the intellectual counterpart of the institutional fusion of science and clericalism in the clerisy. One can extend Heimann's (1972:73) remark that "Though the *Unseen Universe* can be regarded as a popularization of science for an ideological purpose, it was intended as a contribution to the philosophy of nature," by noting that it was both—the latter being the *means* of the former. Analogously, in Oliver Lodge's *Ether and Reality* (1907:179) we find that ether "is the primary instrument of Mind, the vehicle of Soul, the habitation of Spirit. Truly it may be called the living garment of God." The "higher laws" situated in the ether belonged to "a different order of being—an order which dominates the material, while immersed or immanent in it." Nature "must be guided and controlled by some Thought and Purpose, immanent in everything, but revealed only to those with sufficiently awakened perceptions . . . [S]uch reasoned control, by indwelling mind, may be undetectable and inconceivable to a low order of intelligence being totally masked by the material garment." Lodge went on to note the pedagogical and homiletic possibilities of the ether concept:

> The Ether connexion is simpler, more direct, more informing, less dependent on code, more immediately intelligible than anything connected with language. Pictures appeal to children before words. Pictures made appeal to very early humanity before language. Visibile things were apprehended before words. [1907:178]

On the basis of these, and less explicit but nonetheless suggestive statements in Larmor, Rayleigh and indeed several others amongst the Cambridge physicists (cf. Wynne, 1977), it is evident that their science had recognized social value for this group of men. It was a medium of moral demonstration, a symbolic universe of classic conservative reaction (Mannheim, 1938), rooted in the wider intellectual community of the Oxbridge colleges and with natural knowledge centrally embedded within it. To these scientists and intellectuals, science's true role was a broader moral one, in contrast to the utilitarian, antimetaphysical conception of science held by such "professionalizers" as Huxley, Tyndall, Lankester, Carpenter, and Clifford.

The "Cambridge School" inextricably linked physical theory to uses beyond those of the esoteric technical-empirical world, particularly in its integral relationship with psychical research, and more broadly with aspirations to the restoration of a spiritually based social unity to mend the divisive and fragmentary movements in thought and politics associated by the conservatives with the rise of the middle classes. There is a striking analogy between the ineffable, unseen ethereal basis of matter and the articulation of an ineffable spiritual and transcendent basis of social reality—the conservative ideological response to bourgeois individualism. The systemic relational essences underlying material nature symbolized the enduring organic ties of society and acted as the "natural law" witness to a higher spiritual or political law, to be revealed to the ignorant masses by an upper-class intellectual elite. Thus the Cambridge physicists championed the imaginative leap to the "common setting" of thought and the cosmos (Larmor, 1908), and held an a priori sense of the basis of all matter and energy in the superior spiritual realm of ether which ignorant and vulgar materialists could not apprehend.

SCIENTIFIC CONTEXT AND SOCIAL CONTEXT

How to relate the two contexts of use of physical theories and principles is a problem as important as it is difficult. Not the least of the questions it raises is whether two distinct separate contexts can be adequately identified in the first place. But, assuming that they can, the manner of their interaction must be specified as well as its implications for the culture which they shared. On balance, the present case does not appear to be well described as an example of the social exploitation of scientific concepts and theories. Indeed, any presentation which implies one-way traffic from the scientific to the more general context seems inappropriate here. Much more plausible are hypotheses which treat the two contexts as in a symmetrical interaction, or even those which give priority to the general context insofar as the themes and principles discussed in this chapter are concerned. Such hypotheses would imply that features of the general context influenced the cognitive content of late Victorian physics in important and systematic ways. Since this is not the kind of claim which is routinely acceded to at the present time, it is worth reviewing the considerable evidence and powerful arguments which support it.

First, it must be emphasized how radically the late Victorian physicists had transformed their received view of the ether. Doran (1975) has documented the gradual shift from an earlier elastic-solid conception through the late Victorian period, culminating in Larmor's fully electromagnetic, ether field theory of nature. The "ether" remained throughout, but its scientific meaning changed in absolutely fundamental respects. It is no exaggeration to say that "the ether" remained at the centre of a full 180° swing in the conception of matter and nature. Whereas previously matter was the baseline from which ether was a conceptual extension, latterly exactly the reverse was true. At the same time the philosophical status of the ether became increasingly well defined: it was suprasensual but real; ontologically prior to matter, the underlying basis of, and explanation of coherence in, observable phenomena. Thus, for all that they had a received concept of ether, the ether of the Cambridge physicists was essentially their own construction. And it was a construction which other physicists did not deem to be required by the technical state of their discipline.

Secondly, although the constructed ether and associated metaphysics of the "Cambridge School" are difficult to understand entirely in relation to the technical concerns of the esoteric scientific context, they are very readily intelligible in the more general context of use. It is abundantly clear how they functioned in moral discourse, and provided the response to naturalism with an alternative representation of the physical world which appropriately legitimated an alternative social ideal. Moreover, having examined the participation of the physicists in Cambridge intellectual life and particularly in the SPR, there is no difficulty in understanding how the modification of culture in one context could feed back into or interact with usage in the other. Even without the evidence of the more popularizing and polemical writings of the physicists, there is much to show that their natural philosophy was conditioned by broadly based social factors.[4]

Third and finally, if the reality of this connection is accepted, there can be little doubt of its importance as a systematic and continuing factor in the technical context. At the level of theory-construction it is noteworthy that the basic assumption of the ethereal, continuous electromagnetic, nature of matter on the part of the late Victorian physicists also led them to assert the continuous radiation nature of the new emanations, x-rays, gamma-rays, alpha-rays, and beta-rays, which were being found at the turn of the century. This was in sharp contrast to the inheritors of the professionalist mantle, Rutherford and Bragg, who adopted an atomist, particulate paradigm for the full catalogue of emanations (Wynne, 1977; Stuewer, 1971).

Analogously, organic continuity, ontological realism, and the ethereal theory of matter, were put forward as standards of judgment, as the "Cambridge School" expressed their adverse evaluations of the theories of the naturalists and energeticists. One of many good examples is the attack upon the principle of the Conservation of Energy mounted in Stewart and Tait (1874). They argued that the unseen, nonmaterial world of ether, continuous with the world of matter (indeed the constitutive basis of it) and the seat of all energy, must exhibit energy transfers to and from the material realm. Thus, measured energy transformations in the material realm alone would inevitably show discontinuity and nonconservation. Apparent discontinuities in material nature were thus "in reality so many partially concealed avenues leading up the the unseen" ethereal realm (1874: 192). Conservation of Energy would not hold, therefore, in the material world as observed and defined by scientific naturalists such as Huxley, Tyndall, Clifford, and the German energeticists. Against the assertions of these groups that conservation did always apply, the late Victorians argued that the realm of applicability of energy conservation went beyond the confines of the materialist cosmos and incorporated the realm of ether and matter together. Stewart referred to the Conservation of Energy principle as "a weapon against visionaries" who might, like him, be seeking that unseen spiritual realm (Barrett, 1895:309). Scepticism with respect to the Conservation of Energy continued to mark the scientific thought of the late Victorians, as illustrated by Fitzgerald (1896) and Lodge (1913). The physical conception of this matter and ether integration involved a hierarchy of connected levels, with ponderable matter low and vulgar, and ether the repository of a "higher law" in nature. The matter and ether integration, and the ethereal conception of material "strain centres" in the ether, allowed these physicists to put forward the possibility that material particles might dissolve into ether, and vice versa, under certain conditions. This integrated approach to the Conservation of Energy, to the structure of matter, and to matter-radiation interactions, was entirely at odds with the nonether, "corpuscular" approach of the naturalists.

In summary then, the present case appears to be one where the concepts and principles of a science were developed and sustained not only (or perhaps not even) for their technical value, but very much also for their social value. Scientific thought developed in particular ways related to its possible functioning in the general social context rather than the esoteric scientific context. How far general insights can be drawn from this particular case it is difficult to say. Certainly late Victorian science was both less differentiated internally and less divorced from the wider cultural context than science is today. Indeed, the rigidity of these boundaries was one of the issues dividing the "Cambridge" physicists and their antagonists. Apart from the particular interest of these materials, we can suggest that our findings have a bearing on other episodes where scientific concepts are found in employment beyond the esoteric scientific context and where scientific development is impossible to account for on purely "internal" criteria. In such cases there is every reason to look to the general social context to explain the particular course of scientific growth. And in these cases the burden of proof lies firmly on those who reject the social context of use as a formative influence on scientific knowledge.

NOTES

1. As Kargon (1969) observes, Maxwell's own epistemological stance was more conventionalist or pragmatic with respect to the ontological status of scientific concepts than most of his Cambridge peers could accept.

2. Lodge's patronizing attitude and his fear that spiritualists were vulnerable to subversive persuasions is evident from his correspondence, e.g., to J.A. Hill (1932:25).

3. Maurice's attitude towards spiritualist phenomena is indicated in his biography in a letter dated 1870:

> [it] is truth and not fiction, the deliverance from dreams, not the indulgence of them, to hold fast the faith that the veil of flesh has been rent asunder, that for all and for each the invisible world has been opened; that we *must* have converse

with it and its inhabitants whether we desire the converse or shrink from it. [Maurice, 1892:625]

Maurice appears in many ways to have been more ambivalent regarding the ultimate naturalistic basis of progressive social evolution than his friend, Henry Sidgwick, who, whilst loathing the restricted, "vulgar" naturalism of the materialists, nevertheless craved for a naturalistic licence for his own form of "spiritual" upper-class paternalist "progressivism." In this, Sidgwick went much further than true conservatives such as Larmor and Bateson were prepared to go, even though their responses were much the same. The similarities of outlook between Larmor and Bateson are discussed by Wynne (1977), in particular their common abhorrence of reductionism and its social managerial implications. Larmor clearly carried this over into his evaluations as a physicist.

4. Some contradiction is evident between late Victorian intellectuals' profession of a spiritual, organic nature and the utilitarian leanings of some of them, such as Rayleigh and Lodge. However, the structural elements of proper science and its social role which the late Victorian "metaphysical" school articulated can be regarded as stereotypes articulated in *pure*, extreme form in demarcation against the threat from an alternative social group and its ideology. It is probably generally true that a necessary (but not sufficient) condition of the authority of such symbolic universes is that they should be pure and unambiguous, in order to eradicate doubt. In the course of symbolic action between competing social groups, ideologies will inevitably tend to be articulated in forms which are more extreme than the social reality which they are articulated to underpin. Furthermore, to regard apparent contradictions between ideologies, or symbolic universes, and actual social practices as problematic is to imply that the ideologies contain an intrinsic, ideal logic which leads inevitably to the related social practices. Although this is widely held to be so, inside and outside science, it is more valid to see them as weapons in an authority struggle with the ideological justifications of threatening social groups (be they anarchists or industrialists). Thus the social patterns and the symbolic universes so articulated may be genuinely influential upon the nature of natural knowledge developed by each group, without being intrinsic logics of that science, nor the sole determinants thereof. The main elements may be determined less by such supposed logics than by the contingencies of (a) what cultural symbols or parts thereof are available, (b) what would *appear* to be authoritative (i.e., related to established symbols of order, authority, and so on), and (c) what elements would demarcate the groups' ideals more clearly in opposition to the existing symbolic universe of a threatening social group. The relative influence of such ideological determinants and of empirical validity upon the detailed theories of science is something which can only be worked out in specific cases.

REFERENCES

BACKSTROM, P. (1974) Christian Socialism and Cooperation in Victorian England. London: Croom Helm.

BARRETT, W.F. (1895) On the Threshold of the Unseen. London: Routledge & Kegan Paul (reprinted, 1907).

BROWN, A.W. (1973) The Metaphysical Society. New York: Octagon.

BRUSH, S. (1967) "Thermodynamics and history." Graduate J. 7:477-565.

"CAVENDISH" (1910) A History of the Cavendish Laboratory, 1871-1910. London: Longmans Green.

CLIFFORD, W.K. (1875) "The scientific basis of morals." Contemporary Rev. 26:650-662.

CROOKES, W. (1926) Researches into the Phenomena of Spiritualism. London: Psychic Bookshops.

______(1879) "Radiant matter." BAAS Reports: 167.

______(1874) "An inquiry into the phenomena called spiritualism." Quart. J. of Sci. 5:3-19.

d'ALBE, E.F. (1923) The Life and Work of Sir William Crookes. London: Fisher Unwin.

DORAN, B. (1975) "Origins and consolidation of field theory in 19th-century Britain: from mechanical to electromagnetic views of nature." Historical Studies in the Physical Sciences 6:133-260.

DOUGLAS, M. (1975) Implicit Meanings. London: Routledge & Kegan Paul.

FITZGERALD, G.F. (1896) "Ostwald's energetics." Nature 53:441-442.

FLEMING, J.A. (1902) Waves and Ripples in Water, Air and Aether. London: S.P.C.K.

FORMAN, P. (1971) "Weimar culture, causality and quantum theory, 1918-1927." Historical Studies in the Physical Sciences 3:1-115.

GAULD, A. (1968) The Founders of the Society for Psychical Research. London: Routledge & Kegan Paul.

GOLDBERG, S. (1970) "In defence of ether." Historical Studies in the Physical Sciences 2:89-126.

HAINES. G. (1969) Essays on German Influence upon English Education and Science, 1850-1919. Connecticut College: Archon Books.

HEIMANN, P.M. (1972) "The *Unseen Universe*: physics and philosphy of nature in Victorian Britain." British J. for the History of Sci. 6:73-79.

HICKS, W.M. (1895) "Theories of the aether." BAAS Reports:595-606.

HILL, J.A. (1932) Letters from Sir Oliver Lodge, Chiefly on Spiritualism. London: Cassell.

HOWE, E. (1972) The Magicians of the Golden Dawn. London: Routledge & Kegan Paul.

JAMES, G. (1970) Henry Sidgwick. London: Oxford Univ. Press.

JOLLY, W.P. (1974) Sir Oliver Lodge. London: Constable.

KARGON, R. (1969) "Model and analogy in Victorian science." J. of the History of Ideas 30:423-436.
LANKESTER, E.R. (1910) Science from an Easy Chair. London: Methuen.
LARMOR, J. (1908) "On the physical aspect of the atomic theory." Manchester Memoirs in the Physical Sciences 52:1-54.
——(1900) Aether and Matter. Cambridge: Cambridge Univ. Press.
LODGE, O.J. (1913) "Continuity." BAAS Reports:1-47.
——(1907) Ether and Reality. London: Hodder & Stoughton.
——(1889) Modern Views of Electricity. London: Macmillan.
MacLEOD, R.M. (1972) "Resources of science in Victorian England," pp. 172-196 in P. Mathias (ed.) Science and Society, 1600-1900. London: Cambridge Univ. Press.
MANNHEIM, K. (1938) "Conservative thought," pp. 67-131 in Essays on Sociology and Social Psychology. London: Routledge & Kegan Paul.
MAURICE, F.D. (1892) The Life of F.D. Maurice, Told Chiefly from His Letters. London: Macmillan.
——(1886) Social Morality. London: Macmillan.
——(1882) Moral and Metaphysical Philosophy. London: Macmillan.
METCALFE, H. (n.d.) The Evolution of Spiritualism. London: Hutchinson.
MICHELSON, A.A. (1899) "A plea for light waves." Proceedings of the American Association for the Advancement of Science 37:67-78.
NELSON, G.K. (1969) Spiritualism and Society. London: Routledge & Kegan Paul.
NICOL, F. (1972) Essay Review of A. Gauld's, The Founders of the Society for Psychical Research (1968). Proceedings of the Society for Psychical Research 55:341-367.
PODMORE, F. (1902) Modern Spiritualism. London: Methuen.
RAYLEIGH, LORD (1884) "Presidential Address to the BAAS." BAAS Reports:1-32.
ROTHBLATT, S. (1968) Revolution of the Dons. London: Cambridge Univ. Press.
SEELEY, J.R. (1882) Natural Religion. London.
SHAPIN, S. (1979) "Homo phrenologicus," in this volume.
SHAW, N. (1926) "The Cavendish as a factor in a counter-revolution." Nature 118:885.
[STEWART, B. and TAIT, P.G.] (1878) Paradoxical Philosphy. London: Macmillan.
——(1874) The Unseen Universe. London: Macmillan.
STOKES. G.G. (1883) On Light. London: Macmillan.
STUEWER, R.H. (1971) "William H. Bragg's corpuscular theory of x-rays and gamma-rays." British J. for the History of Sci. 5:258-281.
SVIEDRYS, R. (1970) "The rise of physical science at Victorian Cambridge." Historical Studies in the Physical Sciences 2:127-154.
SWENSON, L.S. (1972) The Ethereal Aether. London: Univ. of Texas Press.
THOMSON, J.J. (1936) Recollections and Reflections. London: Bell & Sons.
——(1908) "The late Lord Kelvin." Cambridge Rev. (16 January):5.
TURNER, F.M. (1974a) Between Science and Religion. London: Yale Univ. Press.
——(1974b) "Rainfall, plagues, and the Prince of Wales: a chapter in the conflict of religion and science." J. of British Studies 13:47-65.
WILSON, D. (1971) "The thought of late Victorian physicists." Victorian Studies 23:29-48.
WYNNE, B. (1977) "C.G. Barkla and the J. phenomenon: a case study in the sociology of physics." M. Phil. thesis, University of Edinburgh.
YOUNG, R.M. (1973) "The historiographic and ideological contexts of the nineteenth-century debate about man's place in nature," pp. 344-438 in M. Teich and R.M. Young (eds.) Changing Perspectives in the History of Science. London: Heinemann.

Part Three
THE SOCIAL BASIS OF SCIENTIFIC CONTROVERSY

The subject of the first paper here, by MacKenzie and Barnes, in many ways parallels that of Wynne's essay. The conflict between the Biometricians and the Mendelians has many similarities with that between the scientific naturalists and the late-Victorian Cambridge physicists, and occurred in much the same general social context. However, the present essay gives particular attention to the relationship between judgments and evaluations and the historical settings in which they were made. The primary concern is to illustrate the view that evaluation can never be understood in context-independent terms, and hence that to understand even what might be accepted as "properly scientific" evaluations always requires historical and sociological study of their setting. The judgments of scientists are always historical events occurring in particular historical situations and must be made intelligible in exactly the same way as other such events.

Another, related theme harks back to the arguments of earlier essays. This is the critical appraisal of what might be called idealist forms of explanation, which attribute properties to theories, standards, and representations independent of their use and then invoke the properties as explanations of usage. The argument here is that judgments cannot be explained by the proclaimed scientific standards of those making them, since such standards are themselves actively selected or invoked by those using them. Rather than understanding people as compelled by beliefs and standards, we would be better advised to study them in their particular settings, to try to understand why they accept these beliefs and standards, and employ them in the way that they do. Analogously, it is argued that our sense of the incompatibility, or even the formal inconsistency, of different scientific theories is, in the last analysis, sustained by their relationship to the activities of opposed communities. Ideas, it is suggested, have no inherent properties whatsoever, and have features imputed to them entirely according to their mode of use. Here again the implication is that controversy cannot be successfully studied at the purely formal level, and must always be treated as a concretely situated historical episode.

John Dean's study of alternative methods of botanical classification also serves to illustrate these points. But it has a further, more specific significance in the present context. We have emphasized that anthropological and sociological approaches to scientific culture by no means imply an "externalist" perspective, a point which is now widely recognized largely thanks to T.S. Kuhn's sociological, but "internalist," treatment of scientific subcultures. Nonetheless, the emphasis in earlier contributions has been upon links between the esoteric "scientific" context and the wider culture in which it is situated. That a small community of professional scientists has all the characteristics of a society within itself, to which features of its knowledge can properly be related, has not so far been adequately illustrated. This is a deficiency which Dean's contribution remedies. It takes two modes of botanical classification which have stably coexisted with each other, but have nonetheless been set into opposition and conflict at times, and it relates them to distinct settings and objectives which in both cases would normally be considered part of the esoteric professional context of science. Dean's study might perhaps be called a contribution to "internal" history, in contrast for example to the "external" history in MacKenzie's and Barnes's paper. But the two papers reveal no significant difference in approach or theoretical position. What the terms "internal" and "external" are doing here are pointing to contingent social boundaries, indicating the appropriate sociological reference points to which knowledge is to be related. "Internal" points to the esoteric context; it refers to that within the scientific subculture; it connotes micropolitics, vested interests, the qualified against the layman. "External" points to the wider context beyond the boundary of the scientific subculture; it connotes macropolitics, socioeconomic interests, everyday knowledge—or perhaps other fields of esoteric culture such as religion, art, or political theory. Perhaps there is a useful role of this kind for these terms in the study of the growth of science. But once it is recognized that this is their appropriate use, the "great divide" between the internal history of rational thought and the external history of irrational influences upon it is consigned to oblivion and we are truly beyond the internal/external debate.

The final paper in the volume, by Jonathan Harwood, moves right to the opposite extreme. In terms of our newly created usage, it is a study which relates technical controversy overwhelmingly to "external" factors. Harwood has been studying the current environmentalist/hereditarian debate in the United States for many years now, with particular emphasis upon the race/IQ issue. He has discussed the technical arguments put forward in esoteric professional contexts in a series of earlier papers. But this detailed familiarity with the fine structure of the debate has led him increasingly to consider that central themes within it cannot be understood as attempts to resolve technical problems. In the present paper he suggests that these themes represent alternative available patterns of legitimation, taken up and employed as the general social situation gives opportunity or incentive. Thus, to a large extent the controversy can be seen waxing and waning as general circumstances determine, these circumstances being, according to Harwood, largely economic. Historical contingencies have established a pattern in the United States where increasing expenditure on ameliorative, "Welfare-State" programmes of social control generate demand for environmentalist legitimations, and where constraint upon such programmes consequently encourages hereditarianism, so that controversy between the two positions reaches maximum intensity at predictable turning points in the development of social and economic policy.

8

SCIENTIFIC JUDGMENT: THE BIOMETRY-MENDELISM CONTROVERSY

Donald MacKenzie and Barry Barnes

The study of any particular episode in the history of science involves consideration of a wide range of activities: experiment, manipulation, invention, argument, and judgment must all be investigated in detail in the course of building up an overall reconstruction of what went on. Yet it is sometimes said that scientific judgments and evaluations should not be studied in the normal way, but only in terms of their relationship to the accounts of properly scientific evaluation available in the philosophy of science. Imre Lakatos (1974) has gone so far as to suggest that the history of science should be structured around a "normative methodology" taken from philosophy, and should be explicitly concerned to demonstrate the rationality of science, to the greatest degree possible, as defined by that methodology. It is unlikely that many historians will accept the restrictions that this programme imposes upon their methods and the range of their curiosity. Nonetheless there does appear to be some uncertainty concerning the relationship of abstract philosophical models of scientific judgment and concrete historical investigation.

It is our view that judgments and evaluations should be studied as phenomena like any other, amenable to all the routine general methods of the history of science. Philosophical models of proper evaluation are irrelevant to the historian's task. Indeed, with their typical stress upon the formal, abstract properties of verbal arguments, they can even impede an adequate naturalistic understanding of actual judgments by diverting attention from context and from the goals and interests of particular groups of scientists. General arguments can readily be adduced to support this position.[1] But, here, our primary purpose is to illustrate it with materials relating to one particular historical episode.

BIOMETRY AND MENDELISM

The nature of inheritance and evolution was the subject of continuing controversy, early this century in Britain, between the Biometricians and the early Mendelian geneticists led by William Bateson. This well-known controversy serves as our example. It has been the subject of several historical studies, as a result of which reasonable agreement has emerged upon the sequence of events. We shall thus provide here but the barest outlines of what the controversy involved, and rely upon these earlier accounts to supply further detail, and an appreciation of the historical investigations and interpretations that contribute to its understanding. For references to these accounts and for more detailed discussion and documentation of many of the points made below, see MacKenzie and Barnes (1975) and MacKenzie (1977:246-308).

To understand the differences between the two groups in the controversy one must first understand the extent of their agreement. Both sides accepted and built upon earlier contributions by Francis Galton, whose work established the concept of *heredity* as a relationship between generations (Cowan, 1972). As a result, the first step in the study of heredity became the study of the resemblances and differences between generations, either on an individual or a population basis. Related to his notion of heredity was a distinction between characters received at conception (nature), and those resulting from environmental factors (nurture). Galton argued that the hereditary character of the offspring was a function entirely of the hereditary character of its ancestors; the effects of environmental influences were not transmitted. These notions—the concept of heredity, the nature/nurture distinction, and the rejection of the inheritance of acquired characteristics—provided a common framework for both Biometricians and Mendelians.

Two other themes in Galton's work were, however, very differently perceived by the disputants. One was his claim that statistical methods were crucial to the study of heredity, in justification of which he could point to his own work on regression and correlation. The other was his criticism of the orthodox Darwinian view that evolution proceeded by the selection of small differences, and his suggestion that large variations, or "sports," were essential to the evolutionary process. Biometry took up the first position as its own, but retained the Darwinian view of natural selection. William Bateson, on the other hand, was sceptical of the utility of statistical theory in biology, but was sympathetic to Galton's discontinuous conception of evolution, even before the rediscovery of Mendel's work in 1900 helped him to develop characteristically "Mendelian" methods and theories.

The existence of Biometry as a research school can be roughly dated from 1891, when Karl Pearson and W.F.R. Weldon, both having read Galton's *Natural Inheritance* (1889), initiated a collaboration designed to extend its statistical procedures throughout biology. Since both men occupied chairs at University College, London, they were able to gather around themselves a group of younger researchers to take up the new statistical methods. Biology, and especially evolutionary biology, was to be mathematicized. A comprehensive statement of the Biometricians' position and programme subsequently appeared in one of Pearson's "Mathematical Contributions to the Theory of Evolution" (1896). Here the mathematical tools appropriate to the study of heredity and evolution were displayed: definitions of variation, correlation, natural selection, sexual selection, reproductive selection, heredity and regression were given. And it was made clear that evolution was to be conceptualized in orthodox Darwinian terms, as the consequence of the selection of small, individual differences.

The paper contained a particularly significant discussion of Galton's criticism of orthodox Darwinism and his "Law of Ancestral Heredity." Galton had argued that the selection of small differences could not bring about evolutionary change because of a tendency in offspring to "regress" towards an unchanging species type. Galton's "Ancestral Law" connected the deviation from species type of a given characteristic of an individual to the deviations from type exhibited by all the individuals which made up its total ancestry. Pearson regarded the notion of an "unchanging species type" as a gratuitous assumption and suggested that it was in fact false. The centre towards which regression proceeded was, he argued, the *existing* population mean of the property in question, and thus not an unchanging value but a value which changed as natural selection operated upon the successive generations of a population.

Two extremely interesting consequences followed from this. First, selection of continuous differences could now be very effective indeed, producing a stable new breed in as little as five generations. Thus, the orthodox Darwinian view of selection could be maintained intact and there was no need to postulate "sports" as a cause of evolution. Secondly, the deviations from "type" related together in the "Ancestral Law" all became deviations from the means of real populations. The problematic species type value was replaced with a set of values—all of which had real empirical significance: a population mean was a quantity accessible to measurement. Thus, Pearson could defend the Law in positivistic terms as a condensation of experience based on no theoretical assumptions. Hereditary resemblance could be described by means of a calculus of correlation, and sets of results summarized in terms of the "Ancestral Law."

The definition of heredity used by the Biometricians was first presented in this paper. It was an operational definition:

> Given any organ in a parent and the same or any other organ in its offspring, the mathematical measure of heredity is the correlation of these organs for pairs of parents and offspring. [Pearson, 1896:259]

By this definition all parent-child correlation was ascribed to heredity, thus building hereditarianism into the basis of the overall model of evolution. The definition also led to happy consequences at the technical level, facilitating an easy measure of "the strength of heredity" for a particular character. Its very adequacy in fact made investigation of the *mechanism* of heredity unnecessary to the Biometricians' research programme. In modern parlance, the study of phenotypic resemblances could proceed independently of any consideration of the processes by which these resemblances were produced.

By the late 1890s the Biometricians considered themselves adequately equipped to undertake quantitative studies of heredity and evolution. They possessed statistical techniques for aggregating and processing the measured properties of individuals in large populations, and thus for expressing degrees of resemblance between generations and the "strength of heredity." They could point to what they regarded as particular achievements based on those techniques. And they possessed a general account of evolutionary change, orthodox Darwinism, which both legitimated their techniques and, when "experience" was organized and presented by these techniques, appeared as a "summary of experience." Overall, their methods seemed able to describe degrees of hereditary resemblance between generations, and hence potentially to predict the character of future generations, whether processes of selection were operative or not.

When, after 1900, accounts of heredity were put forward in Britain on the basis of Mendel's rediscovered work, the Biometricians were confronted by a startlingly distinct alternative position. The simple experiments on peas initially cited as examples of the Mendelian approach appeared to indicate a lack of the blending of parental characters assumed by orthodox Darwinism, and to cast doubt on the character of evolution as the product of the selection of small variations. Moreover, as presented by the British Mendelians, these experiments supported a theoretical account of the underlying hereditary processes as constituted by the transmission and segregation of unchanging "Mendelian" factors. Such a theory was, inevitably, more than a "summary of experience" and had initially to coexist with such apparently anomalous phenomena as the "blending" inheritance of height in human populations. And, as a representation of the underlying hereditary process, it offered different predictive possibilities to the Biometric approach. First, like any account of underlying mechanisms in science, it had enormous predictive potential, but little immediate predictive value in everyday contexts. To use a Mendelian scheme predictively one had to make inferences about the Mendelian factors present in the stock from which breeding occurred. This could be done, but it required labourious procedures, and experimentation in breeding, to be worthwhile except in a few fortunate cases; and it certainly was not then apparent how such inferences could be made for existing natural populations. Mendelism, as a predictive resource, appeared initially to imply selected rather than natural populations, controlled settings rather than natural environments, experimentation and manipulative intervention as well as observation and measurement. Secondly, as an account of evolution, or simply of change in the characters of populations, Mendelism was evidently incomplete. As well as accounting for variation in terms of concepts like the segregation and dominance of Mendelian factors, it was clearly essential to account for differences in the overall stock of such characters available in given populations. It was recognized that Mendelism required a hypothesis of "mutation" (in modern terminology). And the ever-present possibility of mutation, of unpredictable change occurring in the Mendelian factors themselves, meant that Mendelism could apparently never generate, even in ideal circumstances, fully secure predictions. On the face of it, Mendelism made evolution unpredictable.

The opposition of the Biometricians to Mendelism stressed these points as its defects, and neither Weldon nor Pearson ever came fully to accept it. Weldon doubted the universal validity of Mendel's laws, even for peas. Green and yellow peas were not completely distinct entities; intermediate colours were present. Further, the evidence suggested that the characteristics of offspring did not conform to the simple patterns reported by Mendel. The theoretical model used by the Mendelians should be replaced by a more flexible one, encompassing both blending and segregating patterns. Pearson was critical of the anti-Darwinian features of Mendelism, and in addition objected to its speculative theoretical character and its endemic weakness as a basis for prediction. These criticisms occur in many passages of his writings, but are brought conveniently together in the following passage from a memoir of 1913:

> The problem of whether philosophical Darwinism is to disappear before a theory which provides nothing but a shuffling of old unit characters varied by the appearance of an unexplained 'fit of mutation' is not the only point at issue in breeding experiments. There is a still graver matter that we face, when we adduce evidence that all characters do not follow Mendelian rules. Mendelism is being applied wholly prematurely to anthropological and social problems in order to deduce rules as to disease and pathological states which have serious social bearing. Thus we are told that mental defect—a wide term which covers more grades even than human albinism—is a 'unit character' and obeys Mendelian rules; and again on the basis of Mendelian theory it is asserted that both normal and abnormal members of insane stocks may without risk to future offspring marry members of healthy stocks. Surely, if science is to be a real help to man in assisting him in a conscious evolution, we must at least avoid spanning the crevasses in our knowledge by such snow-bridges of theory. A careful record of facts will last for ages, but theory is ever in the making or the unmaking, a mere fashion which describes more or less effectually our experience. [Pearson et al., 1913:491]

These critical judgments, however, were of no consequence as far as William Bateson and his supporters were concerned. Bateson had indeed been actively seeking, for some years before 1900, a discontinuous account of evolution to set against orthodox Darwinism. In 1894 he had published a catalogue of *Materials for the Study of Variation* designed to demonstrate empirically that discontinuous variations did occur. And he had included in the work explicit general arguments against the prevalent Darwinist viewpoint:

> We must admit, then, that if the steps by which the divers forms of life have varied from each other have been insensible—if in fact the forms ever made up a continuous series—these forms cannot have been broken into a discontinuous series of groups by a continuous environment, whether acting directly as Lamarck would have, or as selective agent as Darwin would have. [1894:16]

Thus, it was not surprising that Bateson immediately saw a fit between the newly rediscovered work of Mendel and his own ideas. The Mendelian factor, which maintained its identity and refused to blend, was for him clear proof of the discontinuous nature of variation and hence evolution. Bateson and his coworkers went on to take an important role in the development of Mendelian genetics, especially in the first decade of the century—before Morgan's work made the field even more controlled and experimental and sucked it into the laboratory and the fly-room. The word genetics owes its currency to Bateson, and much of the terminology of the subject is his. He demonstrated the existence of Mendelian inheritance in animal species (the original work of Mendel and his rediscoverers was on plants), and did much to elucidate more complex patterns of inheritance than the simple schemes found by Mendel (e.g., the work of Bateson and Punnett on partial coupling).

Over much of the same period Bateson maintained, often polemically, a position of opposition to the work of the Biometricians. He attacked the scope and reliability of the mathematical methods of the opposition as well as their orthodox Darwinism, contrasting their indiscriminate character with the more potent and detailed differentiations which could be made by the trained biologist or naturalist. Bateson's view of evolution involved significant "specific" variations, which contributed to it, and small "normal" variations, which did not, and these were lumped together in the statistical aggregations of the Biometricians (Bateson, 1901). Thus, Pearson's confidence in the predictive reliability of his mathematical methods was not at all shared by his main critic and opponent. The judgments and evaluations of the pair, and of their colleagues and followers, differed over basic points at every level, from that of general theory to that of appropriate technique and what was to count as concrete achievement. And these clashes were never satisfactorily to be resolved, at least insofar as the protagonists themselves were concerned.

CONFLICTING EVALUATIONS

How might the conflict of evaluations all too briefly outlined above be accounted for, if we restrict ourselves to a formal, abstract consideration of the claims and arguments of the two sides? One possibility is to seek opposed criteria, or general standards of evaluation, from which the particular judgments of protagonists can be said to have derived. A conflict can then be understood in context-independent terms as a confrontation of alternative philosophical criteria. Thus, in the present case, it might be suggested that positivist, phenomenalist criteria informed the work of Pearson and the Biometricians, and sustained their opposition to the more theoretical conception of science implied by Mendelism (see Norton, 1975a, 1975b).

To give priority to "criteria" in this way does, however, generate serious problems, not least those of evidence. Even where scientists express and advocate an explicit account of scientific method, as Karl Pearson did, the criteria thereby verbally expressed cannot simply be assumed to have operated in their scientific work. Historians have become accustomed to finding significant gaps between the practice and practical judgments of major figures like Newton, and the general criteria of judgment set out in their writings. And in the case of Karl Pearson it is a simple matter to point to specific procedures and evaluations which fit most uncomfortably with his phenomenalism (MacKenzie, 1978). But even if such gaps were not discernible, it would still be contentious to give priority to criteria. Why should criteria be taken to be the determinants or premises of judgments, rather than rationalizations of them? On what grounds can coexistent historical phenomena be set in an asymmetrical relationship? *Can* a chronological sequence be transformed into a causal one? Certainly, in the case of the present example, it is hard to see what such grounds could be.

There would, in any case, be a further question to be answered, even if a controversy could be made out as a conflict of distinct philosophical standards of judgment. It would remain to be shown why these standards were sustained by the groups of scientists in question. Why should positivism have underlain Biometry, but not the scientific work of the Mendelians? Why should phenomenalist arguments appear powerful and significant to some people but not to others? It might be said that to assert the priority of criteria is merely to reformulate the problem of accounting for clashes of scientific judgment, not to solve it; the procedure leaves the fundamental issues untouched.

There is, however, an alternative approach to these issues, which remains at the formal level, but which is not plagued by the above difficulties. This derives ultimately from the work of T.S. Kuhn (1970), although it might be argued that it misinterprets Kuhn by placing an undue emphasis on the linguistic aspects of scientific inquiry (cf. Toulmin, 1972; Lakatos and Musgrave, 1970). The clash of Biometry and Mendelism, it could be said, was a clash of scientific paradigms. The controversy occurred because the overall linguistic and theoretical frameworks of the two sides were *incommensurable*. Over time the two groups had built up distinct modes of discourse, between which no perfect translation was possible, and through which, accordingly, identical verbally formulated premises of judgment did not extend. Argument could only proceed satisfactorily within the one mode or the other. The controversy was sustained by the fact that the two groups were "trapped" in two incommensurable modes of discourse.

Such a view does not require one component of a system (such as criteria) to be given priority over others, and at the same time it offers a hypothesis accounting for continuing controversy between paradigms. Moreover, it is clear that most of the characteristics of incommensurability described by Kuhn can indeed be found in the present controversy. Karl Pearson (1902:331) himself noted that "Mr. Bateson and I do not use the same language." Let us then briefly examine some of the differences in the language the two sides used.

Significant differences are at once apparent in the explicit definitions employed. In Biometry, the extent of hereditary resemblance, and thus the strength of heredity, was given directly, *by definition*, as the degree of manifest resemblance between "phenotypes" (as we would say). For a Mendelian, on the other hand, such superficial resemblance could only be of significance as an indicator of the operation of Mendelian factors. The latter were the source of all inherited resemblance or variation. Only explanations in terms of these factors ultimately counted as contributions to knowledge about heredity.

Further linguistic differentiation is evident where the two sides applied their theories and methods to particular situations. There was a tendency to label the perceived exceptions to a favoured position "problems," phenomena awaiting investigation and resolution, whereas the exceptions to the opposing position were the evidence against it. Such simple patterns of usage, running along parallel but conflicting lines, did indeed contribute to the formation of alternative inward-looking systems of discourse running right down to the level of "observation-reports" and the description of "data." Moreover, increasingly "theory-laden" language tended to be employed at all levels of discourse as the opposed positions were elaborated. When Weldon, checking Mendel's results by studying the outcome of crosses of different varieties of peas, cast doubt upon the validity of Mendel's laws, he made no impact upon the Mendelians. Bateson responded with the assertion that Weldon's key cases were "mongrel" peas. True Mendelian phenomena would be manifested only by "pure" varieties. But the very notion of a "pure" variety was a theoretical one that presupposed the validity of Mendelian theory (Bateson, 1902:129, throughout).

Thus, very quickly, the two sides evolved distinct modes of discourse which penetrated right into their description of even the simplest and most visible phenomena and which were not mutually translatable. It is clear that, as Kuhn (1970:94) describes it, incommensurability existed between the assertions and arguments of the two sides, and that their differences could not "be unequivocally settled by logic and experiment alone." Nonetheless further consideration makes it apparent that the controversy cannot be explained by incommensurability. The linguistic differentiation indicated above is more the consequence of the two communities continuing to differ than an explanation of that fact. Although the opposed communities did have occasional, temporary problems of communication, overall the evidence suggests that they were able to understand each other remarkably well. Certainly Karl Pearson did not oppose Mendelism out of either ignorance or misunderstanding—quite the contrary. Pearson was thoroughly in command of the language, the mathematical procedures, and the assumptions of the Mendelians. His criticisms were based upon what he knew. Moreover he was no indiscriminate critic. He recognized that in particular instances Mendelian ratios described experience. And he even borrowed and experimented with Mendelian procedures, showing remarkable insight and technical competence when he did so. He demonstrated, for example, that a multifactorial Mendelian model, considered as a model of "phenotypic" resemblance, was essentially isomorphous to his own "Ancestral Law." And, by considering the distribution of Mendelian factors in a randomly breeding population, he arrived at a relationship formally equivalent to the Hardy-Weinberg Law, some years before that law was officially discovered (Pearson, 1904).

Pearson was able to address himself to the Mendelian position and understand it through direct consideration of its problem solutions and its theoretical structure. No process of translation was necessary. Incommensurability is not a significant barrier to understanding for the simple reason that translation is not required for understanding. The typical route to the understanding of culture is that of direct assimilation and active rehearsal of the operations and procedures it includes. Just as Galileo could be a passable Aristotelian, so could Pearson be a more than passable Mendelian. And having assimilated Mendelism, Pearson could, on the one hand, criticize from a Biometric standpoint some of its features, and, on the other, use some of its procedures in a novel way that was more in keeping with the idiom of Biometry.

Thus, the study of incommensurability as a linguistic phenomenon, for all its interest, has not helped us to understand the two alternative patterns of judgment and evaluation. The notion that people are trapped within one pattern or another is untenable. Hence, just as we previously asked why the two sides should have elected to sustain alternative criteria, we must now ask why they elected to sustain alternative paradigms. In our view, this kind of question cannot be answered unless a purely formal, linguistic orientation is abandoned and the alternatives are considered in their actual contexts as forms of culture, coherent patterns of thought and activity, carried on by communities with particular goal-orientations and interests. Such goal-orientations and interests serve to constrain and structure technical commitments and procedural choices within science. Such specific commitments and choices are essential constituents of scientific judgment. So it is not that scientific judgment is on occasion biased by unfortunate intrusions. Rather it is that what we normally accept as properly scientific judgment is always structured and organized by contingent goal-orientations.[2]

NECESSARY CONTINGENCIES

The scientific judgments of the Biometricians and Mendelians were given coherence and structure by the operation of several contingencies of their setting. Only an extended account, based on a wide range of historical materials, would convey even a superficial indication of the general pattern of connections. What follows is no more than a partial view of a complex picture, but it should suffice to illustrate the general points here at issue.

If we seek to describe what it was that the two schools spent most of their time doing, what their activities were mainly concerned to achieve, we arrive at an account reminiscent of Kuhn's (1970) treatment of science as puzzle solving. (Kuhn's work, for all that it has frequently been interpreted as concerned with scientific discourse, is predominantly and most valuably an attempted description of scientific *activity*.) Biometricians and Mendelians alike were oriented to the extension of the competences and procedures characteristic of their different schools. Both communities possessed procedures which, they believed, showed manifest applicability and predictive value in particular modes of use, and both communities tried to extend these procedures and achieve new modes of use by modelling new applications upon existing successful applications.

The Biometricians considered that they had manipulated data on the height of existing human populations so as to achieve a manifestly satisfactory summary of experience with predictive value in the estimation of height in subsequent populations. Their work was devoted to, and organized for, the extension of this and parallel recognized achievements. They sought analogous success with different kinds of data, more troublesome kinds of population, and with particular cases that had initially proved resistant to approved methods and interpretations. And their linguistic practice, as we have seen, mirrored this kind of enterprise so that, for example, "successes" would be presented as evidence for the validity of their approach and "failures" as temporarily unsolved problems. Phenomena were made out in a way that, as it were, oriented them in terms of, and prepared them for treatment by, existing Biometric techniques and interpretations; and this construction upon reality proceeded at every level on the basis of the presupposed validity of the overall approach.

Analogously, the Mendelians possessed a calculus for the manipulation of underlying segregating hereditary factors which they held to yield successful predictions in certain particular crosses employing particular stocks or parents. Their work was directed to the extension of the use of this calculus to further stocks and further kinds of crosses and breeding sequences. And their linguistic practice was structured by this objective, so that the procreation of natural kinds generally was made out as amenable to Mendelian analysis: all hereditary processes were evidence for Mendelism or problems for Mendelism; they were describable by known Mendelian processes, or potentially describable by partly known or even unknown Mendelian processes. It is, of course, the growth of linguistic practice in this way, around distinct alternative procedures and putative techniques of prediction, which gives rise to the formal features of incommensurability when the two positions are considered purely as bodies of discourse.

Concretely considered, the two subcultures, although both engaged in the "scientific" enterprise of extending possibilities for prediction, were pursuing distinct historically specific projects. They sought to extend the scope of alternative predictive techniques and to further explore their capabilities. Success for Biometry would have meant the establishment of a repertoire of potent methods for the prediction of degrees of "phenotypic" resemblance in natural populations. Mathematical techniques were organized and extended to further this objective; phenomenalist epistemology legitimated it. Orthodox Darwinism, which provided grounds for the use of the particular statistical methods involved, would in time have been justified, and revealed as unrestricted in generality, by the demonstrable general applicability of the methods. Success for the Mendelians would have generated an array of competences and predictive methods of a very different kind. At least as perceived at the time, such an array would have had utility and predictive value only in strictly controlled contexts with regard to selected materials—and only then with allowances for inevitable unpredictable episodes. Being a calculus of underlying variables, its employment had necessarily to involve making auxiliary assumptions particular to each instance of use, and attempting to control for all the diverse possible factors that might interfere with the hypothesized mechanism or mask its effects. Mendelism offered predictive resources of great scope and versatility but, initially, of sharply restricted applicability to large natural populations. Furthermore, again as commonly perceived at the time, to establish the general applicability of Mendelian procedures was to undermine orthodox Darwinism and thus to restrict the relevance of the techniques with which Biometry would mathematicize biology. Before 1910, Mendelian procedures themselves required but a minor extension of the existing technical resources of naturalists and biologists.

Thus the activities and the judgments of the two research groups, considered communally rather than as sets of individuals, were goal-oriented. Judgments were made with *particular* sought-after kinds of prediction and extensions of technique being taken into consideration. Distinct specific predictive and manipulative goals prestructured judgment and evaluation. Basically, the two communities were sustained in opposition to each other not by different standards or criteria, nor by problems of translation or intelligibility, but by diverse goal-orientations apparent in theorizing and activity, language and procedure alike.[3]

It might be said that both communities allowed their judgments to be unduly restricted by their peculiarly narrow predictive goals, and that such goals represent pathological departures from a properly open scientific position.[4] But it is misconceived to imagine that some "correct," open, undirected scientific attitude can be specified which is "natural" or "uncaused," and which represents an improvement upon contingently narrowed concerns with prediction. Science is *done*, not merely *thought*, and without any narrowing and particularization of goals, specific activity would become diffuse, incoherent, and noncumulative, as Kuhn (1970) has pointed out. Factors which make for a narrowing and increased specificity of interests in prediction and control thereby make an essential contribution to the research process. But such factors are distressingly contingent and typically sociological. They include esoteric professional interests and associated professional rewards and controls. But they may include as well, or instead, factors outside the immediate culture of professional science.

To recognize science, and scientific judgment, as goal-oriented acknowledges that it might be affected by any number of contextual factors: in terms of idealized conceptions of science, it is exposed to the possibility of pollution. For example, in the present case it seems overwhelmingly likely that the narrowing of predictive goals on the two sides was related to the operation of sociopolitical interests based in the wider culture. Biometry can be linked to such interests in several ways, of which two will be briefly noted. First, the Biometric project was intimately bound up with eugenics, and was explicitly conceived to provide a basis of reliable knowledge for the operation of eugenic interventions. The organization of Biometric research around problems of "phenotypic" resemblance in natural populations effectively centred it upon issues with immediate implications for eugenics. A successful eugenic programme precisely required a reliable measure of "phenotypic" resemblances between generations and an indication of how the modification of population parameters in one generation would affect their value in succeeding generations. Secondly, the Biometric project would vindicate orthodox Darwinism, a doctrine which had at once the character of a scientific theory and an ideological resource, extensively resorted to in political debate and polemic. Karl Pearson himself employed Darwinism both as a theory of evolution and as an account of social change, and frequently appealed to his Darwinist conception of nature to legitimate his analysis of society and his political views:

> You may accept it as a primary law of history, *that no great change ever occurs with a leap;* no great social reconstruction, which will permanently benefit any class of the community, is ever brought about by a revolution. It is the result of a gradual growth, a progressive change, what we term an *evolution*. This is as much a law of history as of nature. [1888:363]
>
> Human progress, like Nature, never leaps. [1888: 122]

This connection does not have to be read in to Pearson. He made it explicit himself:

> The theory of evolution is not merely a passive intellectual view of nature; it applies to man in his communities as it applies to all forms of life. It teaches us the art of living, of building up stable and dominant nations, and it is as important for statesmen and philanthropists in council as for the scientist in his laboratory or the naturalist in the field. [1900: 468]

These connections, with eugenics, and with Darwinism as ideological resource, indicate that the specialization of predictive goals characteristic of Biometry was in the last analysis sustained by social and political interests. The precise character of these interests and the precise way in which they were related to Biometry are issues that need not be gone into in detail here. Elsewhere, MacKenzie (1976, 1977) has suggested that the interests in question were those of the rising professional middle class. New professional occupations sought to benefit from industrialization by making themselves out as purveyors of expertise essential to the control of a changing society. They put forth or supported progressive, gradualist, reformist, and interventionist views of various kinds, and positivistic justifications of expert knowledge. Certainly, their members were richly represented among active supporters of eugenics and proponents of the doctrine of Darwinian gradualism.

It can be argued along similar lines that British Mendelism was related to traditional conservative interests in British society, set in opposition to the ideas of interventionist reformers riding the tide of advancing industrialization. The success of the Mendelian programme would both have added to the predictive resources of biologists and have discredited Darwinian gradualism and certain aspects of the intellectual toolkit of interventionist eugenic reformers. Consideration of the life and work of William Bateson provides some support for this hypothesized connection (Coleman, 1970; MacKenzie, 1977: 286-297).

In these instances, then, there is a case for suggesting that the narrowing and structuring of predictive concerns, and thus of scientific judgment, arose from the operation of broadly based social factors. The case can indeed be extensively argued. But no attempt to do so is made here, since we do not in any case wish to argue for any *necessary* connection with such factors. Particular goals of prediction and control *may be* sustained by esoteric professional interests (such as might once have been called internal factors) and nothing more. And certainly the esoteric interests of professionals are important in understanding scientific judgment. To take an example related to the present case, Allen (1975) has noted how, after Morgan's pioneering work, the scope for the employment of professional skills in the laboratory investigation of inheritance in *Drosophila melanogaster* was one factor which attracted biologists to the Mendelian programme, and thus led them to accept it as a basis for scientific judgment. The general point is not that the goal-oriented character of scientific judgment implies its relationship to any particular contingency, or to external factors, or political interests; what is implied is that any such contingency *may* have a bearing on judgment and that contingent sociological factors of some kind *must* have. What these factors are is always a matter for concrete empirical investigation.

THE RESOLUTION OF CONTROVERSY

We shall conclude our discussion with the briefest possible reference to events subsequent to the controversy. Our purpose is not to provide any further historical materials, but to anticipate an argument. It is well known that R. A. Fisher and others were eventually to incorporate aspects of Darwinism and Mendelism into the new discipline of population genetics, and thus inaugurate a research programme which remains operative today (Provine, 1971). It might perhaps be argued that, in thus transcending and resolving the controversy, work such as Fisher's refutes our general thesis. Science develops and grows. Surely controversies must be resolved, and narrow constraints on judgment transcended, if such development and growth is to occur.

In fact, it is not at all essential for scientific controversy to be resolved if development is to occur. Science is a form of culture like any other. Typically, scientists receive this culture — techniques, procedures, representations, theories — from their scientific predecessors, whereupon they have it available as a set of tools or resources to solve their own problems. Their own solved problems are then fed back into the culture of the research front, to serve as resources in their turn for a further generation of scientists. For this process to occur the resources on the research front need not be in a formally consistent relationship. Nor does the employment of such resources necessarily lead to their being adapted into formal consistency. Scientific work on both sides of a controversy can be taken up for further use and thus contribute to the development of science, but the controversy can nonetheless stand unresolved, a lasting unconformity in the historical record.

Indeed, even this may be an image unduly influenced by formal modes of thought. To talk of consistent and inconsistent relationships between elements of culture, even to assert their irrelevance, may be misleading. We have noted that linguistic patterns on the two sides of our controversy reveal formal incommensurability; this affords us a significant clue. If statements are incommensurable then, ipso facto, they are not formal contradictions. The possibility of making them out as mutually compatible is always there. Such a possibility was indeed explored in the Biometrician-Mendelian controversy, notably by Yule (1902), but, although it echoed themes later to be generally accepted, at the time it was for the most part ignored. We might say that just as the two communities, for reasons already shown, collectively "decided" to remain separate, so they collectively "decided" to define their theories as incompatible. They were not forced to do this by any "inner logic."

Turning to the constraints that narrow and structure scientific judgment, it is true that these change as science develops and grows. But it is quite misconceived to think of their broadening, or weakening, or tending to disappear. Sometimes, when we compare present evaluations with those made earlier in the same field, they have the appearance of being more general and broadly based. But this is because in these cases we have built our later judgments upon the earlier ones and the limitations and restrictions it occurred to us to find in them. If we step aside from this self-justifying perspective, it becomes apparent that present modes of evaluation have no greater claim to complete generality or context-independence than previous modes had. Consideration of the work of R. A. Fisher illustrates this point. His synthesis of Darwinian and Mendelian methods is perfectly intelligible in terms already outlined. His judgment too was prestructured by goals and interests, and he was opposed by those, such as Bateson's collaborator Punnett, who had different goals and interests. (As it happens, Fisher's approach can, like that of Pearson, be related to eugenic and social-Darwinian concerns [Norton, 1978; MacKenzie, 1977: 298-308].)

Thus to move forward from the contingent evaluations of the Biometrician-Mendelian controversy does not allow one to emerge into a period when scientific judgment was "properly" disinterested and context-independent. Scientific research remains, inevitably, patterned by particular goal-orientations whose relationship to the structure of the scientific community and the wider society is always an open question.

NOTES

1. The general argument can be made on the basis of the work of Mary Hesse (1974). Her "network" model of the verbal component of scientific culture illustrates the conventional and endlessly negotiable character of our classifications and knowledge. And it shows that even a concern to increase the overall reliability of our generalizations and laws would still leave an immense range of options in developing our concepts in response to experience. For systematic learning to occur a narrowing of options is necessary: as Hesse puts it, coherence conditions must be set upon the network. It is our claim that coherence conditions are always maintained, the requisite narrowing always produced, by the operation of contingent sociological features of the setting of scientific work. Such contingencies thus make an essential contribution to systematic learning and the growth of knowledge. We regard the work of T. S. Kuhn (1970) as an illustration of a particularly important special case of this general point.

2. Any theory of knowledge which asserts its conventional character tends to lead one on to consider the interests and modes of activity of its users. Thus, in Mary Hesse's account, concepts have no inherent properties of their own; their utility depends upon previous particular instances of their use — it is these instances which make the concepts informative, not their inherent meanings or extensions. Meaning is a matter of how we collectively choose to employ our concepts, not an explanation of how we employ concepts. Analogous conclusions can be drawn from Wittgenstein (1953).

3. It might be agreed that the two positions have to be understood sociologically as alternative communal projects, and yet the stress on goal-orientation might be questioned. Why, it might be said, does one not simply talk of two communities with distinct normative orders and structures of *authority*? This would be close to Kuhn's sociological treatment of paradigm-sharing communities.

In fact, there is no real clash here. The individual would indeed be aware of alternative foci of authority, distinct systems of social control. But one can then ask further why control and authority was applied along the lines that we have described. Patterns of authority do not develop whimsically to sustain any set of random conventions or practices, nor is their stability and continuity self-explanatory. And in seeking to understand and explain these patterns we are led back to communal goal-orientations again.

4. It might even be claimed that the narrowing of the basis of judgment on both sides resulted in a distortion of reality on both sides. It is particularly useful here to treat the two positions as modelled by differently structured Hesse nets (cf. note 1), with the different structures related to different goals and interests. *Hesse nets do not distort reality.* We might even say that reality, with its usual silence and indifference, tolerates both nets alike, just as the surface of the earth will tolerate different maps, or mapping conventions.

REFERENCES

ALLEN, G.E. (1975) Life Science in the Twentieth Century. New York: John Wiley.

BATESON, W. (1902) Mendel's Principles of Heredity: A Defence. Cambridge: Cambridge Univ. Press.

______ (1901) "Heredity, differentiation, and other conceptions of biology: a consideration of Professor Karl Pearson's paper 'On the principle of homotyposis'." Proceedings of the Royal Society 69: 193-205.

______ (1894) *Materials for the Study of Variation.* London: Macmillan.

COLEMAN, W. (1970) "Bateson and chromosomes: conservative thought in science." Centaurus 15: 228-314.

COWAN, R.S. (1972) "Francis Galton's contribution to genetics." J. of the History of Biology 5: 389-412.

GALTON, F. (1889) Natural Inheritance. London: Macmillan.

HESSE, M. (1974) The Structure of Scientific Inference. London: Macmillan.

KUHN, T.S. (1970) The Structure of Scientific Revolutions. Chicago: Univ. of Chicago Press.

LAKATOS, I. (1974) "History of science and its rational reconstructions," Pp. 195-241. in Y. Elkana (ed.), The Interaction Between Science and Philosophy, Atlantic Highlands,N.J. Humanities Press.

______ and A. MUSGRAVE [eds.] (1970) Criticism and the Growth of Knowledge. Cambridge: Cambridge Univ. Press.

MACKENZIE, D. (1978) "Statistical theory and social interests: a case-study." Social Studies of Sci. 8: 35-83.

______ (1977) "The development of statistical theory in Britain 1865-1925: a historical and sociological perspective." Ph.D. dissertation, University of Edinburgh.

______ (1976) "Eugenics in Britain." Social Studies of Sci. 6: 499-532.

______ and S.B. BARNES (1975) "Biometriker versus Mendelianer: eine Kontroverse und ihre Erklärung." *Kö*lner Zeitschrift für Soziologie und Sozialpsychologie, Sonderheft 18 (Wissenschaftssoziologie): 165-196.

NORTON, B.J. (1978) "Fisher and the neo-Darwinian synthesis." Pp. 481-494 in E.G. Forbes (ed.), Hyman Implications of Scientific Advance. Edinburgh: University Press.

——— (1975a) "Biology and philosophy: the methodological foundations of Biometry." J. of the History of Biology 8: 85-93.

——— (1975b) "Metaphysics and population genetics: Karl Pearson and the background to Fisher's multi-factorial theory of inheritance." Annals of Sci. 32: 537-553.

PEARSON, K. (1904) "Mathematical contributions to the theory of evolution, XII: on a generalised theory of alternative inheritance, with special reference to Mendel's laws." Phil. Trans. A, 203: 53-86.

——— (1902) "On the fundamental conceptions of biology." Biometrika 1: 320-344.

——— (1900) The Grammar of Science. London: Black.

——— (1896) "Mathematical contributions to the theory of evolution, III: regression, heredity and panmixia." Phil. Trans. A, 187: 253-318.

——— (1888) The Ethic of Freethought. London: Unwin.

——— E. NETTLESHIP, and C.H. USHER (1913) A Monograph on Albinism in Man, Part 2. Drapers' Company Research Memoirs, Biometric Series 8. London: Dulau.

PROVINE, W. (1971) The Origins of Theoretical Population Genetics. Chicago: Chicago Univ. Press.

TOULMIN, S. (1972) Human Understanding. Oxford: Clarendon Press.

WITTGENSTEIN, L. (1953) Philosophical Investigations. Oxford: Blackwell.

YULE, G.U. (1902) "Mendel's Laws and their probable relations to intraracial heredity." New Phytologist 1: 193-207, 222-238.

9

CONTROVERSY OVER CLASSIFICATION: A CASE STUDY FROM THE HISTORY OF BOTANY

John Dean

Plant taxonomy has traditionally been a descriptive or observational science. This chapter briefly examines attempts made to introduce experimental methods into taxonomy and some of the controversies which have surrounded the use of experimental taxonomic procedures. Experimental taxonomy, or "biosystematics" as it is sometimes called, is a science of fairly recent origin. The possibility of an experimental taxonomy was first outlined in a comprehensive way by the American botanists Hall and Clements (Clements, 1920; Hall and Clements, 1923; Hall, 1929a, 1929b). In the period 1920-1950 the possibility of an experimental taxonomy became the focal point for a great deal of discussion in the taxonomic community. Some of the most important advocates of experimental taxonomy during this time were Göte Turesson, Jens Clausen, W.B. Turrill, John Gregor, and W.H. Camp. Since 1950 biosystematics has continued to be an important element within taxonomic research but, even at the present time, there is by no means universal consensus about the correct role which experiment should play in taxonomic procedure (cf. Davis and Heywood, 1963:451-458). The object of the present essay is to examine the technical and social factors which have maintained and sustained these controversies into the modern period.

This material is of general interest because it lends support to the view that classification is a process of invention rather than discovery, that our classifications of the natural world are "made" rather than "found." If this is the case, then in an important, indeed fundamental, sense, classifications of the natural world have the status of *conventions* and are thus sustained and modified in response to changing patterns of social contingencies.

The further merit of this example is that it lends support to this view in a context which in the past has been thought particularly supportive of the alternative view, namely, that there is a unique pattern of classification isomorphous with the real structure of nature. Anticipating a little, we might say that in the realm of plant classification certain modes of approach have appeared as particularly distinct and obvious, certain ways of drawing boundaries particularly fundamental. However, here, as elsewhere, the natural world is so complex and rich in information that particular ways of drawing boundaries *always* involve selection and processing of information. Particular methods of selection and processing adopted in any one instance are therefore conventional in character, dependent upon prior commitments to certain shared objectives and concerns (Hesse, 1974). To show that *our* classifications of the natural world have the characteristics of invention is strongly to support the notion that *all* systems of classification have such a character. To show that systems of natural classification are designed to maintain and serve shared interests and objectives is to suggest that classification is never a passive and disinterested process of discovery unrelated to social objectives and concerns. Attempts to classify the world of natural kinds have the status of a "hard case" for any theory of classification as invention to overcome, for it is in classifications of this sort that the instrumental and contextual concerns involved in the construction of all systems of classification are least evident and hardest to locate.

The account below is divided into three sections. Section one examines the factors which led to the emergence of experimental taxonomy and the controversies which have surrounded its use. Section two examines an example of the conflicting classifications produced by experimental and orthodox taxonomy. The third and final section deals in further detail with the claim that classification is a process of invention rather than discovery.

THE EMERGENCE OF EXPERIMENTAL METHODS OF TAXONOMY (1900-1960)

A consistent feature of the writings of botanical taxonomists during the current century, and especially since 1920, has been a long and sometimes acrimonious debate over the correct methods and procedures which should be adopted if a successful classification of the plant kingdom is to result. There have been many facets to this discussion and some of these will be discussed more fully elsewhere (Dean, forthcoming), but, broadly speaking, two main bodies of opinion have been expressed. On the one hand, a group of practitioners has emerged which holds that it is essential for correct classification that experimental methods are adopted and that our classification of the living world should be in terms of categories with an experimental base.[1] Alternatively, another large group of practitioners holds that it is external, visual morphological discontinuities of plant form, as perceived by observation rather than experiment, on which classification must depend. Proponents of the latter position represent an approach which is referred to as orthodox, classical, herbarium, or museum taxonomy. Adherents of the former view regard themselves as experimental taxonomists or biosystematists. As Heslop-Harrison (1953:106-121) shows, exponents of each of these two traditions within taxonomy differ significantly in their aims, methods, and theoretical commitments.

Let us briefly illustrate the differences of approach embedded in these two schools of thought and the controversies which have been thereby engendered. For example, Hall and Clements, two early exponents of the experimental approach, argued that it was only by experiment that genuine classification could be achieved.

> [T]he mere recognition of supposed new species in the herbarium hardly merits the term descriptive botany and it can in no-wise be regarded as adequate taxonomic investigation. It has its value, and hence its excuse, in the biological exploration of new and distant countries but, here as elsewhere, permanent taxonomic results must await the application of statistical and experimental methods in the field. . . . Experimental methods promise to turn taxonomy from a field overgrown with personal opinions to one in which scientific proof is supreme. [Hall and Clements, 1923:1]

In contrast, the Dutch botanist Bremekamp maintained that there was no place in taxonomy for experimentation.

> The attitude of the taxonomist towards the progress of genetic investigation should be that of an interested spectator, not more. If he engages himself in hybridization experiments, he should know that he leaves the domain of taxonomy. [Bremekamp, 1939:403]

The highly polemical nature of the debate over experimental methods in taxonomy is well illustrated by some of the replies to Bremekamp.[2] One pro-experimentalist goes so far as to describe Bremekamp's views as "futile" and "a folly," and he concludes his attack on Bremekamp by saying:

> My conception of taxonomy is that of a growing dynamic science, making use of new information and methods as fast as they appear, not that of a hidebound unprogressive static discipline as is usually referred to as closet botany. [Fosberg, 1941:369]

And problems concerning the evaluation of experimental methods in taxonomy have continued from the early part of the present century up to the present (cf. Rollins, 1953).

How did this controversy come into being and by what mechanisms was it sustained? To gain an understanding of the solutions to these problems it is necessary to look at the historical development both of orthodox taxonomy and biosystematics.

Orthodox taxonomy is essentially Linnaean both in its aims and practice. Indeed, the development and long survival of what may loosely and generically be termed Linnaean methods is in itself of great interest and importance. The philosophical presuppositions on which Linnaeus based his taxonomy were probably derived in part from early training in Aristotelian scholasticism. Linnaeus was an essentialist in the Popperian sense of that term (Popper, 1950:31; Stafleu, 1971:25, 31). For Linnaeus species were objective entities, existing in nature and separated from each other by sometimes hidden, but nonetheless real, essences or characters. The taxonomist's purpose was to uncover these real essences and so expose the hidden order of the natural world. The most serious consequence of adopting such a position was that variation *within* the species played a very minor role in Linnaean taxonomy. It is true that the Linnaean system did possess a single category, the variety, which is below the level of the species, but, for Linnaeus, intraspecific variation was a comparatively unimportant matter, a result of cultivation or accident rather than of nature, and he never altered his view that varieties were epiphenomena unworthy of the serious botanist's attention (Stafleu, 1971:90-91).

However, in addition to conforming to these philosophical a prioris, the Linnaean system was also designed to be eminently practical in its application. Apart from an early period in his life, Linnaeus travelled little and his observations of plants relied heavily on herbarium material (Stafleu, 1971:112-114). The Linnaean system, with its ordered hierarchy of class, order, genus, species, and variety, allowed a systematic categorization of plant species based on a visual examination of the plants' gross external morphology such as is possible to perform easily on the dead plant material of the herbarium. In his reliance on the external properties of organisms for constructing a classification, Linnaeus was typical of naturalists of the "classical" period (Foucault, 1970); and in fact this reliance on external features for taxonomy shows up not only in his writings on animals and plants but to a large extent in his writings on mineral classification as well (Albury and Oldroyd, 1977:191-194).

The Linnaean method was designed to allow rapid and easy identification and naming of new plant material. Indeed, Linnaeus himself defined botany as "that part of the natural sciences by which one obtains happily and easily a knowledge of plants and by which one remembers this knowledge" (quoted in Stafleu, 1971:33). Emphasis on the practical, instrumental aspects of classification is obvious in two of the most famous aspects of Linnaean taxonomy: the binary system of nomenclature and the sexual system. Binary nomenclature was, as Linnaeus quickly realized, an ideal system for easily applying a name to a group of plants. "As easily as one names a person," is how Linnaeus himself described its use (quoted in Stafleu, 1971:109). The sexual system, with its emphasis on such easily discernible features of plants as the numbers of stamens and pistils, allowed any new plant species to be rapidly incorporated within the categories of the system.

As Stafleu (1971:143-339) points out in a valuable appraisal of the reception of Linnaean taxonomy, it was the *practical* utility of his methods which so attracted the early botanists and which largely accounts for the phenomenal success of Linnaean taxonomy in the years which followed Linnaeus' death. Descriptions were standardized, names given according to precise rules, a classification was made possible which allowed accurate storage and retrieval of taxonomic information. The practical value of the system was especially evident in late-18th and 19th-century Britain, where it served the needs of an expanding empire in ordering a flood of exotic new plant materials which had scientific, medical, and horticultural importance. In addition, its assumption that species were distinct entities held together by continuous generation and separated by essential differences, accorded well with the Biblical account of species and their origin, an account still prevalent when Linnaeus was writing and a view to which he himself largely subscribed.[3]

The period of some eight decades between the death of Linnaeus and the publication of Darwin's *Origin of Species* saw many developments and refinements in the concepts and practice of taxonomy, albeit within an essentially Linnaean framework. Linnaean nomenclatural reforms eventually gained a nearly universal acceptance and those (e.g., Adanson) who chose not to use them found their work isolated and ignored as a consequence. There was a continuation of the search for a "natural system" of classification and, towards this end, more and more characters, including some internal characters, came to be deployed by taxonomists. In zoological taxonomy larval forms became an increasingly important source of new taxonomic information (Winsor, 1970) and the development of new and better microscopes must have greatly facilitated this process. However, classification continued to be based broadly upon resemblances of form and taxonomists continued to direct their attention primarily to the description of differences between species rather than to an analysis of variation within the species.

By the second half of the 19th century one of the two planks on which the success of Linnaean taxonomy was based had been finally removed from biological thought. If gradual transformation of species occurred, then species could not be the fixed, unchanging entities which they had previously been thought to be. Indeed, if the Darwinian account of gradual speciation was accepted, then the very existence of species as previously conceived was thrown into doubt. Darwin himself subscribed to a nominalistic species concept; for Darwin species were abstractions, fictions of the taxonomist's mind rather than objectively existing entities in nature.[4]

In retrospect, especially if one views progress in science purely in terms of the history of ideas, it is hard to see why the acceptance of Darwinian evolution in the latter half of the 19th century did not precipitate a crisis in taxonomy. In terms of its logic the threat posed by Darwinian natural selection to Linnaean-based taxonomy is not difficult to grasp. If it is accepted that the only basis for a natural classification is evolutionary theory and that species develop gradually, and, if change from one species to another is too gradual to be able to delimit them, clearly species cannot be defined or delineated in the classical manner. As Heslop-Harrison (1953:8-9) says, "the idea of organic evolution was in several ways in conflict with the practices of orthodox taxonomy. Darwin's doctrine was implacably opposed to . . . such an interpretation of natural variation."

However, as a number of later authors have pointed out (e.g., Davis and Heywood, 1963:31-33; Heslop-Harrison, 1953:11), the acceptance of the Darwinian thesis of evolution through natural selection had remarkably little effect on the actual practice of taxonomy during the years which followed its acceptance by the majority of the biological community. The theory of evolution, while it provided a new framework within which the existence of classifications could be interpreted, did not actively provide any new techniques by which such classifications could be produced. The aim of herbarium taxonomy remained to find and describe new species and provide a framework in which the diversity of living organisms could be described and ordered. The existence of such order could now be given an evolutionary or phylogenetic interpretation; organisms were seen as related through descent, but the techniques for producing such classifications remained much as before.

Thus, rather than overthrowing the structure of 19th-century taxonomy, evolutionary theory actually came to provide a source of support for the existing methods. The Linnaean system, with its ordered hierarchy of categories, was seen as a demonstrable proof of the affinities between organisms postulated by the evolutionary hypothesis. It is true that the acceptance of evolutionary speciation cast doubt on the real existence of species as the fixed unchanging entities which had been posited previously, but taxonomists were quite willing to accept that there was a "subjective" element to species-making. To make such an admission was not, in itself, a threat to traditional taxonomic practice. Species might indeed be nominalistic; they might indeed be creations of the taxonomist's mind rather than objective realities, but, if they were a fiction, they were a convenient and indeed necessary one. The writings of the American taxonomist Lynn Bailey (1896) illustrate this point particularly well. Bailey admitted that, given evolution, the species of the early naturalists like Linnaeus were not objective entities but creations of the taxonomist's mind. However, he did not advocate either an abandonment of Linnaean methods or the Linnaean system. Indeed, he argued that it is precisely *because* species are only a convenient fiction that an approach to species-making based on external morphology is to be desired.

Thus, the situation in 1900 was that orthodox taxonomy had seen little change either in its aims or methods since the time of Linnaeus. The aim of classification remained to describe new species on the basis of resemblances of morphology or form, and thus to impose order on the immense variety of natural kinds. The herbarium remained the focal point of taxonomic enquiry. Linnaean categories, with additions where necessary, remained in use and indeed have continued to be employed up until the present day. The Linnaean method, by and large, continued to be perceived as appropriate for the job at hand. As Heslop-Harrison (1953:10) comments:

> during the century-and-a-half after Linnaeus the taxonomic methods which he had so largely helped to establish were successfully employed with little or no modification to complete a major part of what might be termed the primary survey of the higher groups of the plant kingdom. Order was imposed where order had not been before, and to this extent the aims of classical taxonomy were achieved.

What generated the more critical approach to Linnaean taxonomy which has fuelled the controversies of the present century was the development of new theories and techniques not within taxonomy itself, but within at first sight quite unrelated fields. Indeed, it is remarkable that of the early advocates of an experimental taxonomy, almost none were practising taxonomists or members of the established taxonomic community. Hall and Clements, who pioneered the use of experimental methods in the United States, had both obtained Ph.D.s in phytogeography and Clements went on to become a leading figure in the newly emerging discipline of plant ecology. The Swedish botanist Göte Turesson, who formulated the ecotype nomenclature originally as a direct rival to the established Linnaean system (1925, 1930), was also an ecologist by training, and he was working in an institute of plant genetics when his series of classic papers was published during the 1920s. Clausen, who with his coworkers published an enormous literature on experimental taxonomy in the 1930s and 1940s, was primarily a plant geneticist. Gregor, who with Huxley and Gilmour pioneered the cline and deme terminologies, was working in an applied plant breeding institute. Of the early figures in experimental taxonomy, only one, W.B. Turrill at Kew, could fairly be described as an orthodox taxonomist and, even in his case, all his work on experimental systematics was written in conjunction with the cytologist E.M. Marsden-Jones.

Foremost among these developments was the rapid rise of genetics in the years following the rediscovery of Mendel's work in 1900. The synthesis of Darwinian natural selection and Mendelian theories of inheritance during the next two decades paved the way for the emergence of population genetics and, subsequently, for the detailed examination of the mechanisms of speciation and microevolution which characterized the period of the ' new synthesis" during the 1930s and 1940s (Provine, 1971; Adams, 1970; Mayr, 1975). Of great importance also was the recognition of the link between chromosome behaviour and genetics and the subsequent emergence of the new field of cytogenetics.[5]

The new disciplines of genetics and cytology gave to taxonomy a new set of techniques, e.g., transplanting, crossability experiments, detailed chromosome analysis, which allowed an alternative to Linnaean taxonomy to be developed. Also, they resulted in the creation of a community of biologists for whom Linnaean taxonomy, especially in its description of variation at and below the species level, was no longer adequate for the task at hand. Aid in the detailed study of intraspecific variation, rather than identification, became the key concern of the "New Systematics."[6] It was the inadequacy of Linnaean methods to describe intraspecific variation which led botanists like Clausen to urge that "we cannot be content with employing the old methods alone which suited the classification of Linnaean species themselves long ago" (1922:397), and which prompted a later prominent biosystematist to compare Linnaean taxonomy to a "mouldy shroud," inadequate to the task of defining the complex living groups of plant populations as they actually occur in nature (Camp, 1951:127). In effect, the proponents of the new systematics sought to redefine and renegotiate the scope and aims of taxonomic enquiry. Taxonomy was no longer to be seen as mere "museum cataloguing," and the naming of species was to become only the first step of a far-reaching investigation (Mayr, 1942:7).

The response of herbarium taxonomists to the sometimes extremely critical attacks of experimentalists on their methods, procedures, and results has been varied. However, a marked feature of this response is that there has been no wholesale abandonment either of the Linnaean system of nomenclature or of the procedures by which plant species are described, catalogued, and named. The categories of the Linnaean system (class, order, species, and so on) have continued to find employment in monographs and floras; herbarium taxonomists have continued to insist that different species, to be so called, must exhibit clear-cut morphological distinctions which allow their separation from related species.

It is a point of great interest that where orthodox taxonomists have had to justify their continued adherence to traditional methods of taxonomic practice they have usually done so in pragmatic and instrumentalist terms. For classification accurately to reflect the evolutionary or cytogenetic aspects of a plant's history is seen as laudable as an objective, but on practical grounds it is denied that this objective can or should always be acted on.

> Classification should not be inconsistent with what evolutionary evidence is known, although there may be instances where it may be more convenient and serve more purpose if it is. [Davis and Heywood, 1963:xviii]

And again:

> If a general classification is going to be widely used, it needs to work. We must be able to place taxa in higher taxa so that we can find them again. This means that practical considerations, which mitigate against the "objective" aspect of classification, are very important in systematics and necessitate that science shall come to terms with art if biological classification is to be of maximum use to science. [Davis and Heywood, 1963:83]

As a rule, orthodox taxonomists have not denied the value of experimental taxonomy for the production of "special purpose" classifications, but it is denied that such classifications should be seen as a replacement of the existing system. As Cockerell (1926:588) said in an early and generally enthusiastic review of Turesson's work:

> While the ecotype system is highly illuminating, it should not take the place of definite names accompanied by precise descriptions and supported by the type specimens of the herbaria.

And they have pointed to the sheer practical difficulties which stand in the way of the widespread development of experimental methods in taxonomy. As Faegri (1937:401) put it in another review of Turesson's work:

> The thorough taxonomic-genetical analysis of a genus, of its inner structure and relations to other genera . . . is, if not always, the work of a lifetime, at all events the work of many years' intense studies; in the case of trees perhaps the work of centuries. One can hardly expect taxonomists and phytogeographers to wait so long.

The fact that orthodox taxonomists usually legitimate their methods and classifications on pragmatic grounds has resulted in a tendency to see the units of experimental taxonomy as somehow more "objective" or "real" than those of orthodox taxonomy and to perceive the classifications of orthodox taxonomy as having only a heuristic or instrumental value. Certainly there is a widespread tendency for experimental taxonomists to characterize the categories of orthodox taxonomy in this way (Hall and Clements, 1923; Turesson, 1930; Grant, 1957; Dobzhansky, 1958). As Dobzhansky puts it:

> A systematist working with mammals or bird skins or with pinned insects or with dried plants in a herbarium has obviously no direct

> knowledge of whether the forms which he examines could or could not exchange genes. It would be preposterous to expect him to acquire such information before he classifies his specimens. Making a classification cannot be postponed; it is needed now as an aid to all other biological studies. . . . *The species of the systematist are inferences concerning the biological species which are the reality of nature.* [Dobzhansky, 1958:39, emphasis added]

However, it is worth noting that there is no necessity to regard the units of orthodox taxonomy in this fashion. As Davis and Heywood (1963:94-95) point out in an interesting development of an argument first put forward by Valentine and Löve (1958), the units of orthodox taxonomy can be seen as just as "real" as those of biosystematics.

> The taxonomic species is often criticised by biosystematists on the grounds that it is less important biologically than species defined in terms of gene pools and sterility barriers. But . . . this objection can be taken too far, since it would be absurd to underestimate the biological significance of the morphological (and physiological) differences between populations. . . . The form of a plant represents its response to its environment; it has come, in the course of evolution, to have leaves of just such a shape and flowers of just such a colour and to occupy a habitat in which the soil is of a certain texture and acidity. If another plant has leaves, flowers and habitat preferences which are all different, it is arguable that these differences are of at least as great a biological significance as the ability of plants to exchange genes freely. [Davis and Heywood, 1963:94-95]

The continued use of Linnaean taxonomy by orthodox systematists has meant that, for some groups of plants at least, alternative or "competing" orthodox and experimental classifications now exist side-by-side. It is where such competing taxonomies occur that controversies between orthodox and experimental taxonomists have been most acute. This is especially the case at the level of the species, where the conflicting accounts employ different species criteria. For orthodox taxonomists, as we have seen, species are populations of morphologically similar individuals separated by visible morphological gaps from related species. For the biosystematist, species are normally defined as interbreeding populations of individuals, linked by a capacity for gene-exchange. In practice, especially in sexually breeding groups, these two conceptions of the species do not produce conflicting classifications (e.g., all human races are both capable of breeding with each other and morphologically distinguishable from related living primate species). However, frequently, and especially in plants with asexual or partly asexual breeding cycles, this is not the case. It is in these groups that competing taxonomies may be found. The nature of these competing taxonomies can be best illustrated by giving detailed consideration to one botanical example.

CONTROVERSIES IN *GILIA* CLASSIFICATION

The plant genus Gilia (Polemoniaceae) is a group whose biosystematics and genetics have been the subject of a great deal of work in recent decades by Grant and his coworkers in California (for fuller references see Grant, 1957, 1971; and Day, 1965). We shall examine taxonomic problems in two "species" or "species-complexes" within the Gilia genus, these being *Gilia inconspicua* and *Gilia tenuiflora*.

As Day has demonstrated, the Gilia inconspicua complex consists of not less than five interrelated "sibling" or "biological" species. Three of these species are diploid with respect to their chromosome number; the remaining two are tetraploid. Analysis of the chromosome complement of each group shows that one of the tetraploids has been derived, by hybridization and doubling of the chromosomes, from two of the diploids [*G. minor* (2x) + *G. clokeyi* (2x) = *G. transmontana* (4x)]. The other tetraploid is derived also in part from *G. minor* and in part from the third diploid *G. aliquanta* [*G. minor* (2x) + *G. aliquanta* (2x) = *G. malior* (4x)]. Artificial hybridization shows that the five groups, although fertile in themselves, are highly intersterile in all combinations, so that biologically they are good species. However, external phenotypic differences, while present, are not such as to allow identification in every instance. The tetraploid G. transmontana bridges the morphological gap between its two diploid ancestors so that clear-cut morphological distinctions do not occur. Similarly, G. malior intergrades morphologically into its two diploid ancestors. Also, since the G. minor genome is common to both tetraploids, these also cannot in practice always be separated from each other on purely morphological grounds. In practice it is only with the aid of a microscopic examination of the chromosomes that we can distinguish any one of these "species" from the others. Morphologically they form a single group. (The situation is illustrated in Figure 1.)

A second example of a problem group in Gilia concerns the *Gilia tenuiflora-latiflora* complex, a case discussed in detail by Grant (1957). This group consists of at least four different elements generally quite distinct morphologically, but also capable of gene-exchange. Apparently, at some time in the past primary speciation began in this group but was interrupted by subsequent hybridization. (The situation is illustrated in Figure 2.)

Figure 1. Gilia inconspicua complex

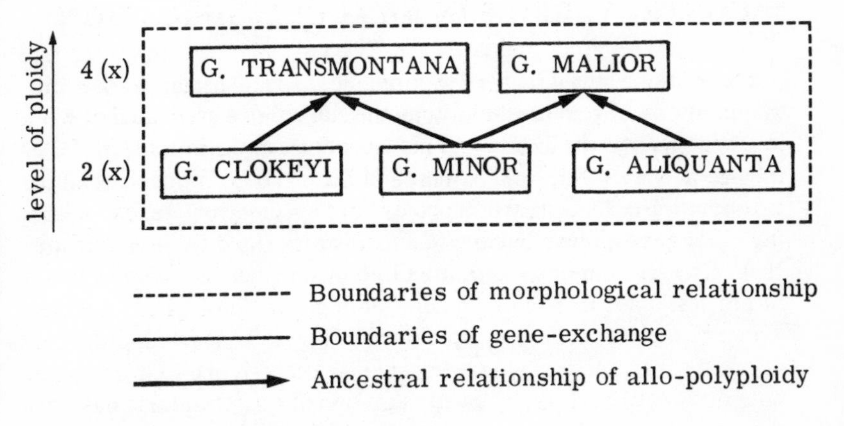

Figure 2. Gilia tenuiflora-latiflora complex

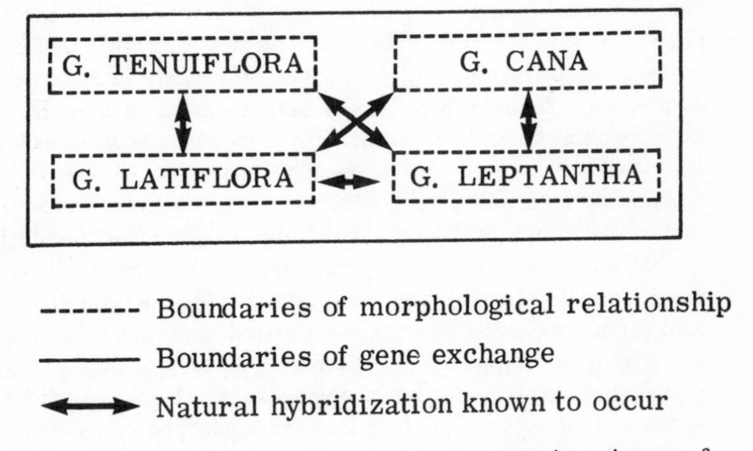

The taxonomic problems here are in many ways a mirror-image of the Gilia inconspicua cases mentioned before. In this instance the rather striking morphological gaps between the constituents of the group would mean that an orthodox systematist would probably wish to retain at least some of these elements as distinct species despite the evidence of gene-exchange. Alternatively, for the strict adherent of the biological species concept, one is dealing here with only a single species. Grant's (1957) discussion of the group recommends that the Linnaean term "species" ought to be abandoned altogether in this particular instance. Instead he uses the term "syngameon" to describe the whole complex and he refers to the individual units as "semispecies." For the formal taxonomist this does not, of course, present a viable solution and such an approach would be at odds with the rules laid down in the International Code of Botanical Nomenclature; for, according to that code, all individual plants belong to a species.

Problems like the ones in Gilia taxonomy outlined above illustrate particularly well the issues which separate orthodox and experimental taxonomy. For the herbarium taxonomist species are units of morphological discontinuity. Practical concerns of identification necessitate that this is the case. As Davis and Heywood (1963:461) put it:

> if we are going to continue to employ the formal taxonomic hierarchy. . . the less we attempt to redefine its categories in evolutionary and genetic terms the better. To do so is to enter such a quaking bog of conflicting aims and interests that practical classification would be paralyzed.

For the experimentalist operating with a species concept based upon capacity for gene-exchange such concerns are not sufficient to prevent splitting, say, a group like Gilia inconspicua into its component biological species. As Grant (1957:57) rather scathingly comments:

> The naming of sibling species as a result of biosystematic studies has provoked a certain amount of discontent among many herbarium curators and floristic taxonomists. The biosystematist does indeed have a responsibility to determine and annotate as many large herbarium collections as is feasible. He also has a responsibility to science not to suppress his findings merely in order to facilitate the task of herbarium filing.

Nor should it be imagined that examples like the ones given here are an exceptional occurrence and hence of little importance to most systematics. Indeed Grant (1957:53) argues that such groups are the rule rather than the exception in higher plant taxonomy.

DISCUSSION

A remarkable feature of the controversy over experimental methods in taxonomy is that the conflict between the two groups of scientists involved has led to no emergent consensus; thus in this respect the present example seems to be at odds with other accounts of scientific conflict in the history of science (cf. Kuhn's [1970] account of conflict and crisis resolution in the physical sciences). On the contrary, questions concerning the value of experimental methods in taxonomy have continued to be voiced for over half a century, and many of the central issues remain quite as controversial today as they were fifty years ago.[7]

This has been the case because alternative classification systems produced by orthodox and experimental taxonomy are *conventions* designed to portray different aspects of reality and to satisfy different demands for technical prediction and control.[8] These different technical goals have given rise to different patterns of scientific judgment and to contrasting classifications of the plant kingdom.

Linnaean taxonomy has continued to flourish into the present century because it continues to serve particular interests in prediction and control not readily replaced by an experimental approach. It makes possible what Heslop-Harrison calls the "primary survey" of the plant kingdom. However, because the method failed to satisfy those biologists concerned with a more detailed analysis of the mechanisms of evolution and speciation, alternatives to the Linnaean approach have been, and are likely to continue to be, formulated. To the extent that nature is conceived of as corresponding to the classification which results, groups of professional scientists have sustained conflicting accounts of what natural kinds exist.

The different goals of the orthodox and experimental taxonomist are pursued in different institutional locations and are thus linked to sets of professional vested interests. Orthodox and experimental taxonomists deploy different craft skills and competences learned in different social networks and communites. As a result, professional objectives and concerns are manifest in conflicts over classification. Concerns of a professional nature are particularly evident in the writings of the experimentalists and these reflect the wider problems of concern to the taxonomic community. Biology in the 20th century is primarily an experimental discipline. Systematics, especially of the more formal kind, tends to be regarded with some disdain and disfavour by the wider community of biologists (Stafleu, 1959; Dobzhansky, 1961). Many taxonomists, and particularly those who are experimentally inclined, have seen in the use of experimental (or sometimes numerical) techniques a means of improving the low status of their discipline (e.g., Constance, 1951:230-231) and have sought to distance themselves from those taxonomists unable or unwilling to adopt the new approach. However, even if we acknowledge the role of narrow professional interests, we are still dealing here with social factors of a kind not easily categorized as "external" to science.

In any event, the present case demonstrates that the alternative classifications of orthodox and experimental systematics are conventions, chosen according to different prior interests in prediction and control and reflecting accordingly the different professional and social interests of the groups concerned. It is also clear that these different interests affected not only which of a given set of classifications was selected or utilized in any given instance, but that they played a vital role in the very *construction* of the classifications. In an important sense the classifications took the form they did *because* of the interests and objectives of those who constructed these classifications.

As a philosophical problem the argument between realism and instrumentalism has endless ramifications and appears to be doomed to indefinite continuation. Nor can every component of realist accounts of classification as discovery be dismissed even at the level of naturalistic description. However, at this level, the present example does concern classifications which in many important respects do have a conventional, instrumental character and it indicates how, in these respects, classifications generally might be expected to have such a character.

Interesting problems arise when we attempt to evaluate competing taxonomies such as those produced in the Gilia examples. If classification is a process of discovery, then a direct appeal to nature should be sufficient to evaluate the best taxonomy. Is this in fact the case? Evidently, no. In this case both taxonomies are conventions designed to emphasize different aspects of the real world. Consider again Gilia inconspicua: morphologically, we do indeed find only a single species; experimentally, evidence can be accumulated by which five different species may be discerned. Both taxonomies are built upon perceptible, systematizable, stable distinctions between individual plants. In this sense the natural order sustains both taxonomies; neither can be said to involve a distortion of the real facts; neither can be said to be erroneous. Nature does not in itself allow such an evaluation to be made.

In conclusion then, it has been argued here that in many important ways classification is best seen as a process of invention rather than discovery. A consequence of this invention model is that classifications can be seen as conventions which are maintained, sustained, and modified in the light of social interests and concerns. Using the example of experimental systematics in botanical classifications, it is possible to locate what some of these changing social interests and concerns have been and to elucidate their respective roles. It is a point of some importance in that none of these interests were in any way scientifically inappropriate or in some way "beyond the bounds" of science. In the traditional idiom of the history of science, the controversy between orthodox and experimental taxonomists has been "internal" to science. And yet, as has been shown, the controversy remains intelligible only if science is considered as human activity and if reference to such factors as socialization, patterns of reward, objectives, and interests is taken into account. These are sociological issues. To deploy sociological explanation is not necessarily to do "external" history of science or to show the influence of "external" factors on scientific growth. Even in a case like this, without "external" sociopolitical factors being particularly evident, a sociological approach is essential to an understanding of the development and distribution of different classifications of the natural world.

NOTES

1. The actual kinds of experimental techniques employed have varied. Probably most important have been studies of crossability, such studies being designed to examine the limits of gene-exchange between populations either in nature or in the laboratory. Of great importance also have been transplant experiments designed to examine the differences between phenotypic and genotypic variations in plant populations. Experimental taxonomists have also been much involved in studying chromosomes and their role in plant evolution. For a modern account of the methods of biosystematics see Solbrig (1970).

2. For the full account of the clash between Bremekamp and Fosberg and the context in which this debate emerged see Bremekamp, 1939, 1942; Fosberg, 1939, 1941; Uittien, 1939; Gilmour and Turrill, 1941).

3. Towards the end of his life Linnaeus altered his view on the origins of species and claimed that nature blended the genera, thereby producing species. However, as Stafleu points out, the overall framework of Linnaeus' thinking remains, even here, creationist and there is no justification in seeing this later work as a forerunner of transformism (Stafleu, 1971:134-139).

4. Interestingly, one of the results of the New Systematics has been to reaffirm the objectivity of the species. For Darwin species were convenient fictions; for the modern geneticist they are a reflection of real barriers in nature to gene exchange (Darwin, 1859:484-485; Mayr, 1955).

5. The importance of cytogenetics to the growth of experimental taxonomy and neo-Darwinian evolution is a field which appears to have been comparatively neglected by historians of science, although a very brief account of the discipline's developments can be found in Sturtevant (1965:33-38).

6. The term "New Systematics" derives from Huxley (1940).

7. Perhaps the true analogue with physics in the present example is to compare orthodox and experimental taxonomy with the roles allotted to Modern and Newtonian mechanics. As Hesse (1974:61-66) shows, there is a formal incommensurability in the

two mechanics, just as the species concepts of experimental and orthodox taxonomy are also generally incommensurable. But this formal incommensurability does not mean that, in certain contexts, both mechanics may not still find employment. For many practical purposes classical mechanics still provides a suitable framework for the investigation of problems (e.g., in engineering). For the more esoteric concerns of the academic physicist modern mechanics is often a more appropriate resource. The difference, of course, is that the ontological priority of the New Mechanics is not in doubt while, as we have seen, the ontological priority of biosystematic categories and classifications is still a matter of dispute.

8. The notion of "interests in technical prediction and control" stems initially from the work of Habermas but is used here in the sense discussed by Barnes (1977:1-26).

REFERENCES

ADAMS, M.B. (1970) "Towards a synthesis: population concepts in Russian evolutionary thought, 1925-1935. J. of the History of Biology 3:107-129.

ALBURY, W.R. and OLDROYD, D.R. (1977) "From Renaissance mineral studies to historical geology in the light of Michel Foucault's *The Order of Things.*" British J. for the History of Sci. 10:187-215.

BAILEY, L.H. (1896) "The philosphy of species-making." Botanical Gazette 22:454-462.

BARNES, B. (1977) Interests and the Growth of Knowledge. London: Routledge & Kegan Paul.

BREMEKAMP, C.E.B. (1942) "Controversial questions in taxonomy." Chronica Botanica 7:255-258.

_____(1939) "Phylogenetic interpretations and genetic concepts in taxonomy." Chronica Botanica 5:398-403.

CAMP, W.H. (1951) "Biosystematy." Brittonia 7:113-127.

CLAUSEN, J. (1922) "Studies on the collective species of *Viola tricolor* L. II" Botanisk Tidsskrift 37:363-416.

CLEMENTS, F.E. (1920) "Ecology." Carnegie Institution of Washington Year Book 19:341-366.

COCKERELL, T.D.A. (1926) "Ecotypes of plants." Nature 117:588.

CONSTANCE, L. (1951) "The versatile taxonomist." Brittonia 7:225-231.

DARWIN, C. (1859) The Origin of Species. London: John Murray.

DAVIS, P.H. and HEYWOOD, V.M. (1963) Principles of Angiosperm Taxonomy. Edinburgh: Oliver & Boyd.

DAY, A. (1965) "The evolution of a pair of sibling allotetraploid species of cobwebby Gilias (Polemoniaceae)." Aliso 6:25-75.

DEAN, J.P. (forthcoming) "Genecology and the historical origins of the new systematics 1900-1953." Ph.D. dissertation, Edinburgh University.

DOBZHANSKY, T. (1961) "Taxonomy, molecular biology and the peck order." Evolution 15:263-264.

_____(1958) "Species after Darwin," pp. 19-55 in S.A. Barnett (ed.) A Century of Darwin. London: Mercury Books.

FAEGRI, K. (1937) "Some fundamental problems of taxonomy and phylogenetics." Botanical Rev. 3:400-423.

FOSBERG, F.R. (1941) "For an open-minded taxonomy." Chronica Botanica 6:368-370.

_____(1939) "Taxonomy and hybridism." Chronica Botanica 5:397-398.

FOUCAULT, M. (1970) The Order of Things: An Archaeology of the Human Sciences. London: Tavistock.

GILMOUR, J.S.L. and TURRILL, W.B. (1941) "The aim and scope of taxonomy." Chronica Botanica 6:217-219.

GRANT, V. (1971) Plant Speciation. New York: Columbia Univ. Press.

_____(1957) "The plant species in theory and practice," pp. 39-80 in E. Mayr (ed.) The Species Problem. Washington, D.C.: American Association for the Advancement of Science.

HALL, H.M. (1929a) "The taxonomic treatment of units smaller than species." Proceedings of the International Congress of Plant Science, Ithaca, New York, 1926, 2:1461-1468.

_____(1929b) "Significance of taxonomic units and their natural basis . . . from the point of view of taxonomy." Proceedings of the International Congress of Plant Science, Ithaca, New York, 1926, 2:1571-1574.

_____and CLEMENTS, F.E. (1923) The Phylogenetic Method in Taxonomy. Washington, DC: Carnegie Institute.

HESLOP-HARRISON, J. (1953) New Concepts in Flowering Plant Taxonomy. London: Heinemann.

HESSE, M. (1974) The Structure of Scientific Inference. London: Macmillan.

HUXLEY, J. (1940) The New Systematics. Oxford: Clarendon Press.

KUHN, T.S. (1970) The Structure of Scientific Revolutions. Chicago: Univ. of Chicago Press.

MAYR, E. (1973) "The recent historiography of genetics." J. of the History of Biology 6:125-154.

_____(1955) "Species concepts and definitions," pp. 1-22 in E. Mayr (ed.) The Species Problem. Washington, DC: American Association for the Advancement of Science.

_____(1942) Systematics and the Origin of Species. New York: Columbia Univ. Press.

POPPER, K. (1950) The Open Society and its Enemies: Vol. I, The Spell of Plato. London: Routledge & Kegan Paul.

PROVINE, W.B. (1971) The Origins of Theoretical Population Genetics. Chicago: Univ. of Chicago Press.

ROLLINS, R.C. (1953) "Plant taxonomy today." Systematic Zoologist 2:180-190.

SOLBRIG, O. (1970) Principles and Methods of Plant Biosystematics. London: Macmillan.

STAFLEU, F.A. (1971) Linnaeus and the Linnaeans. Utrecht: A Oosthoek's Uitgeversmaatschappij N.V.

_____(1959) "The present status of plant taxonomy." Systematic Zoologist 8:59-68.

STURTEVANT, A. (1965) A History of Genetics. New York: Harper & Row.

TURESSON, G. (1930) "Genecological units and their classificatory value." Svensk Botanisk Tiddsskrift 24:511-518.

_____(1925) "The plant species in relation to habitat and climate: contributions to the knowledge of genecological units." Hereditas 6:147-234.

UITTIEN, H. (1939) "Reflections on the nomenclature of so-called hybrids." Chronica Botanica 5:212-214.

VALENTINE, D.H. and LÖVE, A. (1958) "Taxonomic and biosystematic categories." Brittonia 10:153-166.

WINSOR, M.P. (1969) "Barnacle larvae in the nineteenth century: a case study in taxonomic theory." J. of the History of Medicine 24:294-309.

10

HEREDITY, ENVIRONMENT, AND THE LEGITIMATION OF SOCIAL POLICY

Jonathan Harwood

> Any policy implies the reasons by which it could be refuted. In appealing to the values which justify it, it must disparage others which are also valid. . . . To . . . dramatize its own necessity, reform seems to proceed most characteristically by polarizing the issue and insisting upon the side of the debate least honoured in the prevailing order. . . . Because of this, the movement of reform tends to be circular, continually redressing the balance by returning to preoccupations against which the last reform was itself a reaction.
>
> —P. Marris and M. Rein, 1967: 236

The relative importance of heredity and environment in the causation of individual and group differences in socially important traits (especially intelligence test performance or 'IQ') is an issue which has long occupied American academics. The first extended clash between "hereditarians" and "environmentalists" dates from the 1920s, and there have been periodic eruptions of controversy ever since. The most recent episode was triggered in 1969 with Arthur Jensen's (1969, 1973) provocative claims about the neglected importance of heredity for racial differences in IQ.

It is significant that while hereditarian interpretations of race differences in IQ had previously appeared in 1958 (Shuey), 1961 (Garrett), 1964 (Ingle), and 1966 (Shuey), these knowledge claims (though criticized at the time) failed to produce either the sheer volume or the ferocity of the lay and academic response to Jensen et al. since

1969. Since Jensen's formulation of the hereditarian case is undeniably more detailed and sophisticated than his predecessors', it is difficult to see how an "intellectualist" approach can adequately explain scientific change in this instance. The contextual model I outline below regards "controversiality" as historical: differences of theoretical perspective are necessary but not sufficient for scientific controversy. Theorists can happily maintain opposing views in a state of peaceful coexistence as long as such theories remain inconsequential for groups' practice. Whenever professional or political interests are at stake, however, theoretical differences are transformed into heated controversy and elements of the intellectual substance of those theories will mirror threatened interests (cf. MacKenzie, 1978; Shapin, 1978; Nowotny, 1975).

We may now ask whether the protagonists in the most recent episode of race-IQ controversy differ in their political outlooks. In earlier analyses (Harwood, 1976, 1977) I concluded, like Pastore (1949) before me, that environmentalists tend to be left of hereditarians. Nevertheless, it is important not to exaggerate the breadth of this political gap; the liberal political assumptions *shared* by Jensen's supporters and (most of) his critics are probably more important than their differences. As I will argue elsewhere (Harwood, forthcoming), such slight political differences as exist are insufficient for explaining various central features of the controversy. Instead of looking only for features of actors' social situations (typically membership of social class or other groups) which "push" them into opposite sides of scientific controversy, we must also look for macrostructural features of the larger society which "pull" actors from largely similar social situations into widely divergent intellectual positions. In the theory of heredity-environment controversy discussed below I suggest which features of social change in the United States since the 1950s created competing *legitimatory roles* to which Jensen's supporters and critics were differentially recruited. But where do such roles originate?

THE WELFARE STATE AND THE POLITICS OF TINKERING

Historians of industrial society have charted some of the paths by which the modern welfare state emerged from the laissez-faire capitalism of the 19th century. They have tried to show how bursts of legislative fervour represented the response by ruling classes to a problem of order (cf. Goldthorpe, 1964). Studies of the "Progressive

Era" have shown how American industrial leaders at the turn of this century, faced with recurrent economic recession, national strikes, impoverished slums increasingly populated by immigrants, and the rapid growth of trade unions and the socialist movement, welcomed a broad progressive reform movement which could stabilize an increasingly precarious capitalist society without seriously disturbing the distribution of power (Kolko, 1963; Weinstein, 1968; Wiebe, 1967; Hays, 1964). Much the same argument has been made for Roosevelt's New Deal reforms of the 1930s (Bernstein, 1970).

As ruling classes are periodically confronted with problems of control, the welfare state's legislative apparatus expands or contracts to manage disruptive groups, alter unemployment levels and the like (cf. Piven and Cloward, 1972). Such policy shifts must be justified in terms of their likely *efficacy* in solving the social problems which are their target. Doubters of conservative inclination must be persuaded that human weakness can be counteracted through intelligent policy design. Those with radical doubts must be persuaded that major structural change is unnecessary to solve the problem. In the Progressive Era the main source of reformist optimism was "scientific method." Progressive intellectuals expressed their faith in man's ability to render serious problems tractable through rational analysis, experiment, and the creation of efficient new institutions; "social engineering" was the guarantor of progress (Kaplan, 1956; Rogin, 1967). Many progressives, furthermore, saw education as the most effective agent of social control (Karier, 1975; Karier et al., 1973; Mills, 1966).

The legitimation of welfare state policy shifts can also be made in terms of hereditarian or environmentalist theories. The former's stress on individual and group differences can be used to extend the differentiation of the labour force and the educational system or to justify the contraction (usually) of welfare provision. Environmentalist theory can be invoked when marginal groups need to be integrated into the labour force, to reduce the degree of differentiation in the educational system, or to justify the extension (usually) of welfare provision. Russell Marks (1975) has used this idea to develop a theory of heredity-environment controversy. He argues that the Carnegie Corporation's decision in the 1930s to commission Gunnar Myrdal's (1944) research for *An American Dilemma: The Negro Problem and Modern Democracy* was prompted by corporate capital's growing belief that Negro labour would have to be more widely exploited in order to take up the slack left by declining immigration. Justifying his criticism of discrimination, Myrdal endorsed an environmentalist theory of racial differences in mental ability (in contrast with the prevailing hereditarianism of the 1920s) and warned of the dangers of Negro discontent for America's domestic and foreign relations. After the War, Marks argues, continuing rapid economic growth remained a central factor in the extension of civil rights legislation in the early 1950s. Once again environmentalist theory was used to justify this extension, and the liberal social psychologists from the Society for the Psychological Study of Social Issues, whose testimony weighed heavily in the Supreme Court decisions of that time, have figured prominently amongst Jensen's critics in the 1970s (Harwood, 1977). Thus "nurture," no less than "nature," can be of legitimatory use to the power behind the welfare state.

But why are these legitimatory roles so attractive to academics? Ladd and Lipset (1975) have concluded that throughout the 20th century American intellectuals have been strongly left-liberal (social scientists more strongly than other disciplinary groups). Their support for "progressive" social reform rather than radical change is probably only due in part to the social stratum from which they are recruited. The other relevant feature of their social location is their *professional role*. Much of the welfare state's appeal for academics lies in its apparently *technocratic* structure; it is a political order which appears to value expertise greatly. As Chomsky (1969:273) observes:

> the welfare state technician finds justification for his special and prominent social status in his "science," specifically in the claim that social science can supply a technology of social tinkering on a domestic or international scale.

Throughout the 20th century reform movements such as "Progressivism" and the "New Deal" which appeared to allocate a large role to expertise have been warmly supported by American academics (Ladd and Lipset, 1975; Lipset and Basu, 1976).

I will argue below that antipoverty reforms from the early 1960s created a legitimatory role which proved attractive to American academics, especially to social scientists anxious to demonstrate the relevance of "environmentalist" theory to human betterment. By the late 1960s this reform programme had lost its utility for the ruling class and needed replacing. The subsequent policy shift in turn created a legitimatory role for theorists who could explain the failure of the 1960s reforms and justify the reduction of public spending on further reform. This role offered professional opportunities to those academics able and willing to occupy it (such as Jensen and his supporters).

Let us now put this theory of the race-IQ controversy to work.

THREE CENTRAL FEATURES OF THE CONTROVERSY

In previous analyses various features of the controversy's cognitive dimension have either been overlooked (Urbach, 1974a, 1974b) or inadequately explained (Harwood, 1976, 1977).

(a) Despite the enormous variety of intellectual positions taken by academics on the race-IQ issue (Harwood, 1976:385), the controversy is polarized to a considerable extent into supporters and critics of Jensen who typically refer to one another as "hereditarians" and "environmentalists," respectively. Within each of these groups there is a tendency for members to adopt characteristic clusters of belief on separate, technical issues. For example, Jensen's supporters believe not only that the race-IQ difference is substantially genetic but also that (1) high heritability estimates for IQ are correct and (2) IQ is an adequate measure of intelligence. Jensen's critics, on the other hand, tend to accept lower heritability estimates for IQ and to dispute IQ's equation with "intelligence."

Although it is logically possible for interrace differences in IQ to be entirely *environmental* in origin while heritability estimates are *high* within either race (Lewontin, 1970), Jensen's critics rarely adopt this position. Conversely, although it would be consistent for interrace differences to be entirely *genetic* while heritability estimates were *low*, Jensen's supporters have not been receptive to claims of low heritability estimates. Similarly, while Jensen's critics could in principle have granted IQ as an adequate measure of intelligence, they have tended not to do so. And although Jensen's supporters could logically have doubted the equation of IQ with intelligence, they have instead been staunch defenders of IQ's adequacy. That the latter are in no way compelled to take up this position, however, can be seen in the fact that the occasional radical critique (e.g., Jencks et al., 1972) accepts the contention that race differences might be genetic while persuasively arguing that IQ correlates poorly with traditional criteria of intelligence. Thus there appear to be no narrowly "intellectual" grounds for these belief-clusters; an explanation for them will be found in the political significance attributed to heritability and IQ by supporters and critics (see also Harwood, 1977:3-5).

(b) On the whole, therefore, the debate seems to revolve around heredity and environment. Nevertheless, when Jensen's supporters and critics are dichotomized in this way, a small number of interesting anomalies emerge. For example, several of Jensen's critics, while they have taken exception to various technical features of the hereditarian argument, do *not* in fact dispute the plausibility (or even the likelihood) of genetically based IQ differences (Bodmer and Cavalli-Sforza, 1970; Hirsch, 1971; Dobzhansky, 1973). Conversely, one of Jensen's strongest supporters, while attributing social class differences in IQ to genetic factors, does not believe that one need invoke genetic factors in order to explain racial differences in IQ (Herrnstein, 1971, 1973). Furthermore Herrnstein greatly admires E.C. Banfield's analysis of urban blight in *The Unheavenly City* (1968) in which Banfield explicitly denies that blacks' social problems have genetic origins. Instead poverty and disadvantage are seen as the product of inadequate and intractable values: "lower-class culture." By situating the controversy in its social context, I will show how Jensen's anomalous hereditarian critics and environmentalist supporters can be explained.

(c) In an earlier analysis (1976), I concluded that Jensen's critics and supporters differ not only in the intellectual *content* of their positions (e.g., their assessment of the theories' and data's adequacy), but in the *style* in which their positions are expressed. Whereas Jensen's supporters' position is characteristically rationalist, quantitative, abstract, atomistic, and static, his critics' position is inclined to be intuitive, qualitative, concrete, holistic, and dynamic. These two constellations have, of course, frequently been juxtaposed in Western cultures, especially since the 18th century (e.g., Williams, 1958; Nisbet, 1966). In his analysis of one such juxtaposition in early 19th-century Germany, Mannheim (1953a, 1953b) argues that these two styles derive from contrasting conceptions of society. The former style is to be found amongst thinkers whose conception of society is structured by the metaphor of the *machine*. The latter style is derived from an *organismic* conception of society and the thinkers' quest for the creation or maintenance of community (*Gemeinschaft*). The divisive thrust of the former style can be contrasted with the integrative thrust of the latter.

This integrative thrust can be clearly seen in the views of Jensen's critics. Several of them have been actively involved in promoting racial integration in the United States (e.g., Pettigrew, Gordon, Mercer, Deutsch). Others have argued on genetic and evolutionary grounds the *unity* of the human species against those who defend the division of the species into races (Harwood, 1977:13). Many of Jensen's critics stress the importance of compensatory education programmes to bring minorities back into the "mainstream of American life" (Harwood, 1977:13-14, 28-29).

Jensen's supporters, in contrast, share a highly individualistic model of society in which any forms of social policy which are directed towards particular *groups* are an anathema (cf. Harwood, 1976:380):

> You are not your race; you are not your group. You are you. That is, if you're talking genetics. . . . Genetics in fact contradicts this kind of typology which compels so many persons to identify with various groups as if the statistical attributes of the group determined their own characteristics. [Jensen, 1973:10]

Jensen (1973:10) derives a meritocratic moral of a highly atomistic kind:

> Racism and social elitism fundamentally arise from identification of individuals with their genetic ancestry; they ignore individuality in favor of group characteristics; they emphasize pride in group characteristics, not individual accomplishment; they are more concerned with . . . quotas than with respecting the characteristics of individuals in their own right. This kind of thinking . . . is anti-Mendelian.

Since mechanistic and organismic styles are not dictated by the substance of the hereditarian and environmentalist positions, respectively (Harwood, 1976:393), the origin of these styles must be sought elsewhere.

Here, then, are three features of the controversy which, while puzzling in the abstract, become intelligible in the context of American social policy since the 1960s.

THE PROBLEM OF ORDER IN 1960

During and after the War the mechanization of Southern agriculture displaced several million Negro sharecroppers who came to Northern and Western cities in search of work. Rapid automation of manufacturing industry in the 1950s, however, substantially reduced the demand for unskilled labour, upon which urban Negroes had traditionally relied (Baran and Sweezy, 1968). Further aggravated by unions' and employers' discriminatory practices, Negro unemployment rose sharply, as did rates of juvenile delinquency. Gunnar Myrdal observed in 1958:

> In spite of the very rapid advances made towards national integration, heterogeneous elements still linger everywhere in the population, and with the remnants of separatistic allegiances. [cited by Moynihan, 1969:161]

Giant American foundations were evidently concerned. The Carnegie Corporation financed James Bryant Conant's study of the American high school in the course of which Conant warned of "allowing social dynamite to accumulate in our large cities":

> The building up of a mass of unemployed and frustrated Negro youth in congested areas of a city is a social phenomenon that may be compared to the piling up of inflammable material in an empty building. . . . Potentialities for trouble—indeed possibilities of disaster—are surely there. [1961:18]

Like Myrdal before him, Conant warned of the disadvantages of this situation for America's foreign relations and recommended increased federal spending on slum schools to equalize educational opportunity for Negro children.

The failure of urban renewal programs to alleviate this situation plus the ineffectual and conservative character of the federal government prompted the Ford Foundation in the late 1950s to initiate the "Grey Areas Projects." These attempted to develop the schools of several large Northern cities as relevant community centres to meet educational and other needs of slum children and adults (Marris and Rein, 1967:10-20). Furthermore the trustees of the Foundation encouraged its planners to address the problem of delinquency, and the latter duly produced the first of the "community action programs" of the 1960s: "Mobilization for Youth" (MFY) sought to reduce delinquency through extending educational and vocational opportunity to slum youth (Moynihan, 1969:54-55).

> The process of assimilation into the mainstream of society had broken down, and the remedy must lie in the reopening of opportunity. [Marris and Rein, 1967:20)]

By proposing slum adults' participation in the planning and administration of MFY, the Foundation hoped to foster a

> sense of identification with the community and the larger social order. People who identify with their neighborhood and share common values are more likely to control juvenile misbehavior. [Moynihan, 1969:106-107]

Some historians of this period have attributed the strong support for antipoverty legislation in the 1960s to "the rediscovery of poverty" in the 1950s (e.g., Meranto, 1967). Yet as Levitan (1969) shows, the economy was booming in the early 1960s and there was very little party political or popular interest in this "rediscovery." Similarly the Civil Rights Movement itself seems to have made little impact on poverty legislation until the massive March on Washington in August 1963, when it began to shed some of its preoccupation with *civil* rights in *Southern* contexts (Moynihan, 1969). It was only after the summer of 1963, when black civil rights demonstrators replied with uncharacteristic violence to police brutality (the Birmingham, Alabama riots), that congressmen and the Cabinet began to take an interest in the problem of poverty. In October, President Kennedy's Council of Economic Advisors' concern about the continued failure of economic growth to lower Negro unemployment and their recommendation of a "Concerted Assault on Poverty" began to find a warmer reception (Moynihan, 1969:79-80).

Later that winter, President Johnson's advisors were faced with the task of devising a legislative programme for the spring of 1964 which would ensure his reelection in November. Although existing community action programmes were, to a considerable extent, experimental and had not yet proven their practical efficacy, Ford Foundation planners had been working very closely with federal agencies since 1962 (Marris and Rein, 1967; Moynihan, 1969), and their influence on the first "War on Poverty" (WP) legislation drafted in the White House at this time was substantial. Modelled on Ford Foundation programmes, the WP stressed expanded opportunity for the poor via (a) job-training schemes for the "hard-core unemployed," (b) supplementary ("compensatory") education for the disadvantaged, and (c) "community action agencies" to organize the urban poor. In the rush for an immediate election pay-off, however, there was no time for carefully designed, theory-based programmes to be advanced and tested on an experimental basis as MFY had been (Marris and Rein, 1967:210-212).

> The President wanted action, not planning; wanted nationwide scope, not target areas; wanted in particular to see that Negroes got something fast, without in the process alarming whites. [Moynihan, 1969:84]

Like its predecessors, the WP was bathed in the rhetoric of "the quest for community" and "consensus" (Wiebe, 1975; Moynihan, 1969; Alinksy, 1965; Marris and Rein, 1967:43-54). Johnson expressed this mood in his 1964 State of the Union Message:

> the central problem is to protect and restore man's satisfaction in belonging to a community where he can find security and significance. [Moynihan, 1969:80-81]

Even critics noticed this preoccupation of the WP and conceded that perhaps the WP was necessary

> as a symbol of national identity and of that mutuality of concern and interest without which government would be naked coercion. [Banfield, 1969:257]

In the summer of 1964 the WP bill quickly became law. The Ford Foundation had been extraordinarily effective in moulding government policy and towards the end of 1964 could begin to withdraw its funding for community action programmes (Marris and Rein, 1967:29).

ENVIRONMENTALISM'S ROLE IN THE WAR ON POVERTY

The WP's programmes, like its Ford Foundation predecessors, were justified by reference to environmentalist theories: undesirable behaviour was the product of inadequate environments rather than innate incompetence. Furthermore, these environmental determinants were thought to be alterable through relatively modest institutional change. Mobilization for Youth had tried to reduce delinquency through job-training and supplementary education. The Grey Areas Project, too, had seen education as the point of attack in breaking the "cycle of poverty." Compensatory education for preschool children (e.g., Project Headstart) was supported by the view that certain minorities' intellectual and educational disadvantages are the result of various environmental handicaps, among them discrimination (Morton and Watson, 1971; Ginsburg, 1972). Office of Education and Department of Labor pamphlets strongly endorsed an environmentalist interpretation of race differences in IQ (Harwood, 1979).

In view of the fact that President Kennedy evidently had read Hunt's influential environmentalist book, *Intelligence and Experience* (1961), one might wonder whether environmentalist theory in the late 1950s and early 1960s encouraged the reform optimism of the WP. This is highly unlikely on several grounds. First, in his book Hunt (1961:82) had warned that the environmentalist research which he was reviewing "[was] still a challenge for investigation rather than for application in social change." Second, the environmentalist theories of deviance and poverty upon which the WP was predicated did not enjoy unqualified support among academic specialists; had the planners conducted even a brief and superficial survey of relevant social science opinion, they would have discovered alternative environmentalist theories which were far less optimistic about the ease of reform (e.g., theories which saw poverty and deviance as reinforced by "lower class culture" and thus very difficult to eradicate [Moynihan, 1969:170; cf. 1968]). Furthermore, well before the WP's demise, several prominent educational psychologists who had endorsed compensatory education were beginning to complain that environmentalist theory had been "oversold" and that some government agencies had totally unrealistic expectations about how quickly educational competence could be increased (Harwood, 1979). It seems clear, therefore, that environmentalist theory's role in the WP was not that of stimulus but of *rationalization*.

Whatever reservations they may have had about the planners' enthusiasm for their theories, several social scientists became committed to the WP's compensatory educational programme as consultants, chairmen of White House committees, or directors of WP-funded research on compensatory education (Harwood, 1977). Since Jensen's (1969) monograph opened with the charge that compensatory education had failed, it is hardly surprising that these social scientists soon joined the ranks of his most prominent critics.

But why did policy makers seize upon *environmentalist* theory to legitimate the WP? Could not a reform programme have been justified on the grounds that as a group Negroes were unfairly handicapped *genetically* in the meritocratic struggle for success and thus deserved extra help from WP measures until the genetic handicap could be alleviated? Some genetic diseases are, after all, very simply cured while various individual differences shaped through the environment (e.g., personality) are exceptionally difficult to change. In principle this should have been possible; in practice it was not:

> A society's legitimations must necessarily be understood by reference to its previously given culture, and hence its *history*, as well as the immediate context in which the legitimations are set forth. [Barnes, 1977:84]

Academic heredity-environment debate was presumably chosen as a source of legitimatory principles because of its longstanding association with American social reform. Throughout the 20th century hereditarian thought had been employed to justify policies which, to modern eyes, appear regressive: immigration restriction, antidemocratic policies, racial segregation, and so on. Environmentalist thought during this period acquired distinctly progressive connotations through its history as a justificatory principle. Nazi eugenics and Stalinist Lysenkoism forcefully strengthened these connotations. Given these traditional meanings for heredity and environment, the WP planners naturally opted for the latter.[1]

THE DECLINE OF THE WAR ON POVERTY AND THE RETURN OF HEREDITARIANISM

The WP quickly ran into opposition from big city mayors insisting on their right to control WP programmes through municipal bureaucracies. Unwilling to forego the support of local Democratic parties who dominated big city politics, the White House yielded (Rose, 1972; Yette, 1971; Marris and Rein, 1967). The escalation of race riots from 1964 despite the WP began to encourage its critics, and in the face of rapidly growing expenditure for the Vietnam war, the Administration was in no position to protest Congress's annual cuts in WP appropriations (Moynihan, 1969). By late 1966 the WP was becoming a political liability for Johnson and the Democratic Party. In 1968 Moynihan concluded his analysis of the WP, commenting:

> The 1960s, which began with such spendid promise of new and higher unity for the nation, are ending in an atmosphere of disunity and distrust of the most ominous quality. [1969:202-203]

With the WP nearing conclusion by the summer of 1968, federal social policy had begun to shift gears and the environmentalist optimism of the WP came under attack. In September of 1968 Arthur Jensen began to draft his provocative monograph (1969). Concluding his history of 19th-century American school reform, Michael Katz wrote prophetically in 1968 that

> in the past few years education has been infused with an element of zeal; it has once again become a cause. . . . Even if disappointing, it will be fascinating to watch the movement unfold. . . . In the 1830s and 1840s reform started in a passionate blaze of optimism resting on the assumption that environment has prime influence in forming mind and character. The beginning of disenchantment in the [1860s] was signalled by the appearance of theories stressing the importance of heredity. . . . Once again [in the 1960s] environment has become the rage, the commitment of reformers. If it passes again from fashion, if we view a strong and pervasive reassertion of hereditary stress, then we shall know that this reform movement has gone the course of its predecessors. [1968:216]

If the foregoing reconstruction of American social policy in the 1960s is correct, several otherwise puzzling features of the race-IQ controversy become intelligible. The race-IQ issue was *relatively* uncontroversial between 1958 and 1968 because environmentalist theory was endorsed by the ruling class in order to legitimate the extension of welfare state policy. As I argue elsewhere (Harwood,

1979, and forthcoming), American academics (and especially social scientists) generally supported the progressive social reform of this period and largely took for granted the validity of its associated environmentalism.[2] As long as reform proceeded satisfactorily and enhanced the prestige of the social sciences upon whose knowledge it was ostensibly based, there was little reason to get excited about hereditarian challenges.

Given environmentalist theory's pride of place in the WP, however, Jensen's 1969 monograph could only be seen as politically regressive and professionally damaging by those social sciences which had served the WP. This time hereditarianism posed a substantial threat which could no longer be dismissed. Outside the social sciences, though professional practice may have "pushed" them closer to hereditarian positions, a handful of geneticists (e.g., Dobzhansky, Hirsch) could nevertheless agree with the *politics* of the WP. After 1969, though their positions on the race-IQ issue were sometimes only marginally different from Jensen's, they joined their social science colleagues in attacking Jensenism and its apparent policy implications. Now that the WP is being rejected as a hopelessly optimistic relic of the past (Miller, 1978), there are signs that antihereditarian criticism is declining (Harwood, 1979).

That hereditarianism should burst upon the American scene once again in 1969 is understandable in terms of the structure of opportunities presented to academics by the shift in American social policy. That sector of the academic community which patiently harboured unpopular views during the 1960s now seized the opportunity to advance their professional interests. For Cattell, Eysenck, and Ingle this meant expressing hereditarian views which they had long held, while for Jensen it meant adopting such views for the first time (Harwood, 1976). For others such as Herrnstein and Banfield, it meant advocating a position which, though *formally environmentalist*, was just as pessimistic about the WP's impact. Within the discipline of psychology, behaviourists and other scientifically oriented psychologists went into battle against their clinically oriented colleagues (Harwood, 1976). Despite these differences in intellectual positions (as well as considerable differences in personal political views), all of these academics were drawn into the same legitimatory role. The abandonment of the WP required scientific justification,

> And for any particular action experts can certainly be found in the universities who will solemnly testify as to its appropriateness and realism. [Chomsky, 1969:252]

HEREDITY-ENVIRONMENT CONTROVERSY AND STRATEGIES FOR SOCIAL CONTROL

The mechanistic style of contemporary hereditarianism is no doubt in part the consequence of the abstract, static, atomistic, and quantitative character of the biometric and psychometric research traditions from which Jensen et al. constructed their positions. Similarly the organismic style of environmentalist thought is probably due in part to the character of the particular cultural resources (e.g., Piaget's model of intellectual development) assembled by environmentalists. Nevertheless, it is unlikely that the divisive thrust of hereditarian thought and the integrative thrust of environmentalist thought is explicable solely on these grounds. In addition we should think of hereditarian and environmentalist thought as incorporating imagery from the competing strategies for social control which they legitimate.

In his study of American IQ controversy in the 1920s and 1930s Russell Marks (1972) suggests that underlying the heredity-environment dispute was a clash between two strategies for maintaining the meritocratic social order. Hereditarians put forward a *divisive* strategy in which IQ testing would ensure a stable but divided social order by fitting individuals into jobs for which they were ideally suited. Opposing this, the environmentalists' strategy stressed *integration*. The similarities among men were emphasized, and men such as Walter Lippmann, William Bagley, and John Dewey stressed the need for shared values and the minimization of social conflict. "Instead of intensifying biological differentiation," Bagley argued, we should use education to "stimulate cultural integration" (Marks, 1975:324). In the educational sphere the hereditarians advocated streaming while their opponents stressed "common aspirations" and the "cultivation of common traditions" which would foster the mutual understanding, sympathy, and cooperation deemed crucial to a democratic society. As antistreamers saw it,

> resemblances [between persons] in ideas, ideals, aspirations and standards . . . unite men by bonds that are vastly stronger than are the differences in native endowment that would otherwise pull them apart. [Marks, 1972:163]

Many Progressives, too, had been critical of the hereditarianism of their time: Dewey, Bagley, and Lippmann as well as C.H. Cooley, Frank Freeman, and Lester Ward (Pastore, 1949). Furthermore, the Progressives sought an anti-individualistic alternative to Spencerian laissez-faire. Their quest for community has been well-documented (Karier et al., 1973; Mills, 1966; Hofstadter, 1955; Wiebe, 1967;

Noble, 1959), and for Dewey and Addams—as for their modern counterparts—this quest for community centred on the problem of assimilating marginal ethnic groups into the mainstream of American culture. As Rogin (1967:193-194) has argued, the Progressives

> feared the society would fragment, that basic conflict between the new poor and the new rich would split it apart. . . . But harmony could be created from above by a neutral, administrative state. The state could eliminate conflict, or at least limit it. . . . Reform was seen as the outcome not of conflict but of consensus management.

The parallels with the communitarian thrust of the WP and modern environmentalism could hardly be clearer.[3] The strategies for social control implied in the writing of Jensen's critics and his supporters differ in two basic respects. For the critics: (1) the threat to political stability lies in urban racial unrest caused by local whites' discriminatory practices. (2) Black separatism and militancy must be undermined by absorbing the black community into the mainstream of American society through the extension of equal opportunity, principally via education. For Jensen's supporters, in contrast: (1) The threat to political stability of urban racial unrest had been aggravated by inflated Negro expectations. (2) These now needed to be carefully deflated; Negro problems must no longer be portrayed as readily solvable (Banfield, 1968:259-261; Harwood, 1977:18). In a memorandum to the President early in 1969 Moynihan raised the question, "What has been pulling [American society] apart?" and concluded that it was "the Negro revolution" and the war in Vietnam. To deal with the former problem, he advised Nixon to "de-escalate the rhetoric of crisis about the internal state of society" (Yette, 1971:86-88). The threat of growing Negro *group* consciousness is countered by insisting that the black population is simply a statistical aggregate of *individuals* widely differing in genetic endowment. Policies which are tailored for them *as a group* thus fly in the face of the facts (Harwood, 1976:379-381). Negro disadvantage is seen not as a problem deriving from *group*-based discrimination but as the sum total of *individuals'* incompetence.

Of course to refer to mere differences of "strategy" is to imply that, like their counterparts in the political sphere, the intellectuals supporting and attacking Jensen do not disagree over political fundamentals. As I argue in detail elsewhere (Harwood, forthcoming), it is a mistake to portray as "reactionary" such proponents of contemporary hereditarianism as Jensen, Herrnstein, or Eysenck. They see themselves, justifiably, as modern "liberals"; theirs is the politics of the broad centre, overlapping the moderate wings of both Democratic and

Republican parties. Similarly the vast majority of Jensen's critics basically share his supporters' liberal meritocratic political outlook (Harwood, 1977:17-19). The difference between them lies in the critics' doubts that equality of opportunity has yet been achieved as adequately as Jensen's supporters tend to believe. Most importantly, both groups (in contrast to certain radical critics such as Richard Lewontin, Christopher Jencks, or Noam Chomsky) regard poverty as the product of mismanagement which can be rectified through modest adjustments of social policy within the existing economic system. As Eysenck states:

> [our] criticism is not of their motives, which are identical with ours; it is of their lack of concern for facts which must decide . . . what is the best course to be pursued. . . . The argument is about means, not about ends. . . . We all want to help the disadvantaged; our disagreement is on the best way to do it. [Eysenck, 1977:96; cf. 1971:152]

The belief-clusters described above now begin to make sense. As meritocrats, Jensen's supporters and critics alike see the ideal society as allocating rank on grounds of a measure of intellect which would (a) correlate perfectly with the individual's social performance and (b) differ from person to person solely on genetic grounds (heritability = 100%). Jensen's supporters see each of these criteria as largely attained; thus (a) IQ is an adequate measure of merit and (b) heritability estimates for it are high. His critics regret the imperfect functioning of the meritocratic order: (a) IQ is not quite good enough and (b) heritability estimates are not as high as Jensen claims. The consistency within supporters' and critics' belief-clusters is thus not logical but contextual in origin (see also Harwood, 1977:4).

CONCLUSION

The theory of heredity-environment controversy discussed here raises a great number of tasks for further contextual analysis, many of which should be easier once more historians have addressed themselves to postwar America. We need to know, for example, which social structural developments prompted the ruling class to deploy an integrative strategy in the early 1960s followed by a divisive one in the late 1960s.[4] One focus of this research should certainly be the role of corporate capital via the foundations in sponsoring innovative programmes to the point where governmental bodies can be persuaded to incorporate them into official policy. This appears to have been the Ford Foundation's stategy from the late 1950s. After 1966 the

Foundation played a major role in promoting private business investment in the ghetto, encouraging decentralization of New York City's schools (despite local white protest) and financing almost all the main Negro protest groups (Allen, 1969).

It must be obvious that the Foundation's community action programmes were never intended to eliminate American poverty. One need only note the programmes' neglect of the aged and rural poor (Levitan, 1969), the planners' apparent naiveté in hoping that they could persuade local bureaucracies that it was "rational" to relinquish control of welfare measures (Rose, 1972; Marris and Rein, 1967), and the programmes' provision of job-*training* but not jobs. It is more plausible to see community action programmes as a policy designed by a progressive, national sector of the ruling stratum which attempted to reestablish a social order which had been disturbed by the impact of postwar economic change on the black minority.[5] The *local* sector of this ruling stratum, faced with burgeoning Northern ghettoes but a declining tax base, tried unsuccessfully to contain black unrest through traditional strategies of social control: police, schools, welfare agencies. National corporate capital then, with its broader perspective and superior resources, attempted to intervene in order to liberalize the strategies of the local sector. Unable, apparently, to bring about such a change by *direct* negotiations with the local sector, the national sector devised the community action strategy in order to organize increasingly militant blacks for protest through conventional political channels, thereby *indirectly* inducing local politicians and bureaucracies to become more flexible and innovative in quelling the grievances of such potentially disruptive groups (cf. Marris and Rein, 1967; Piven and Cloward, 1972; Levitan, 1969:153; Allen, 1969:70; Wiebe, 1967, 1975). In time, however, the local sector's resistance forced the abandonment of integrative strategies emanating from Washington and a return to decentralized decision-making after 1968 (Wiebe, 1975:113; Moynihan, 1969:102-104, 180-181). Whether current shapers of American educational and social policy are deploying hereditarian theory as part of a new strategy for social control remains to be seen.

NOTES

1. It has often been remarked that proven technical efficacy was not a prime criterion in the planners' choice of a theoretical basis for WP programmes (Marris and Rein, 1967; Moynihan, 1969). On the other hand it is conceivable that the planners selected fairly carefully from the relevant social sciences those theories whose organismic style would reinforce the WP's communitarian intention. It would be worth

considering the extent to which certain areas of environmentalist theory would have been *technically* adequate for the legitimatory task but too "hard" or mechanistic in *style* (cf. Chein, 1972; Hudson, 1972).

2. Banfield (1968) and Moynihan (1969) have suggested that the WP attracted particular support from the well-meaning, upper middle-class, liberal intellectuals who tend to occupy the prestigious universities, the Ford Foundation, the White House, and the quality national media and who have little knowledge of, or sympathy with, the problems of the white working- and lower-middle classes. Why such a group might have been attracted to the WP, its communitarian rhetoric, or its environmentalist legitimation deserves further study.

3. The theory advanced here, however, predicts that environmentalist thought will be couched in a markedly organismic style where ruling classes deploy such thought to *reintegrate* the social order. Should such a class find some variant of environmentalism useful in legitimating a *divisive* strategy, however, one would expect environmentalist theory to acquire more mechanistic features.

4. One theory relevant to this issue is that in times of rapid economic expansion, increased differentiation of the workforce (in order to divide and rule) is facilitated through increased differentiation of educational institutions (Bowles and Gintis, 1976). At first glance, however, economic expansion in America 1900-1920 and throughout most of the 1960s seems to be associated with reform movements with *integrative* intentions. Similarly the divisive strategy of post-1968 hereditarianism has reemerged in a period of economic stagnation.

5. The White House's and Congress's interest in the WP may have owed more to the pressures of the 1964 election than to the long-range concern of corporate capital with urban social control. Piven and Cloward (1972) argue that Democratic Party leaders like Kennedy and Johnson, faced with growing black electorates but indifferent white local leadership in the Northern cities, saw the community action feature of the WP as a relatively cheap way to help urban blacks organize in order to get their share of the resources of local welfare agencies. This pay-off would hopefully attract the new black electorate to the Democratic Party.

REFERENCES

ALINSKY, S. (1965) "The war on poverty—political pornography." J. of Social Issues 21:41-47.

ALLEN, R.L. (1969) Black Awakening in Capitalist America. Garden City, N.Y.: Doubleday.

BANFIELD, E.C. (1968) The Unheavenly City. Boston: Little-Brown.

BARAN, P. and SWEEZY, P. (1968) Monopoly Capital. Harmondsworth: Penguin.

BARNES, B. (1977) Interests and the Growth of Knowledge. London: Routledge & Kegan Paul.

BERNSTEIN, B.J. (1970) "The New Deal: the conservative achievements of liberal reform," in B.J. Bernstein (ed.) Towards a New Past: Dissenting Essays in American History. London: Chatto & Windus.

BODMER, W. and CAVALLI-SFORZA, L.L. (1970) "Intelligence and race." Scientific Amer. 223:19-29.

BOWLES, S. and GINTIS, H. (1976) Schooling in Capitalist America. London: Routledge & Kegan Paul.

CHEIN, I. (1972) The Science of Behaviour and the Image of Man. London: Tavistock.

CHOMSKY, N. (1969) American Power and the New Mandarins. Harmondsworth: Penguin.

CONANT, J.B. (1961) Slums and Suburbs. New York: McGraw-Hill.

DOBZHANSKY, T. (1973) Genetic Diversity and Human Equality. New York: Basic Books.

EYSENCK, H.J. (1977) "When is discrimination?" pp. 93-98 in C.B. Cox and R. Boyson (eds.) Black Paper 1977. London: Temple Smith.

_____(1971) Race, Intelligence and Education. London: Temple Smith.

GARRETT, H. (1961) "The equalitarian dogma." Perspectives in Biology and Medicine 4:480-484.

GINSBURG, H. (1972) The Myth of Cultural Deprivation. Englewood Cliffs, N.J.: Prentice-Hall.

GOLDTHORPE, J.H. (1964) "The development of social policy in England, 1800-1914." Trans. 5th World Congress of Sociology 4:41-56.

HARWOOD, J. (forthcoming) "Heredity, environment and the lure of technocracy."

_____(1979) "Credibility and social change: the race-intelligence controversy in the United States since 1960." (Typescript available from author)

_____(1977) "The race-intelligence controversy: a sociological approach II—external factors." Social Studies of Sci. 7:1-30.

_____(1976) "The race-intelligence controversy: a sociological approach I—professional factors." Social Studies of Sci. 6:369-394.

HAYS, S.P. (1964) "The politics of reform of municipal government in the Progressive Era." Pacific Northwest Q. 55:168.

HERRNSTEIN, R.J. (1973) IQ in the Meritocracy. London: Allen Lane.
______(1971) "IQ," Atlantic Monthly, September.
HIRSCH, J. (1971) "Behavior—genetic analysis and its biosocial consequences," in R. Cancro (ed.) Intelligence: Genetic and Environmental Influences. New York and London: Grune & Stratton.
HOFSTADTER, R. (1955) Social Darwinism in American Thought. Boston: Beacon Press.
HUDSON, L. (1972) The Cult of the Fact. London: Jonathan Cape.
HUNT, J.M. (1961) Intelligence and Experience. New York: Ronald Press.
INGLE, D. (1964) "Racial differences and the future." Sci. 146 (15 October):375-379.
JENCKS, C. et al. (1972) Inequality: A Reassessment of the Effect of Family and Schooling in America. New York: Basic Books.
JENSEN, A.R. (1973) Educability and Group Differences. London: Methuen.
______(1969) "How much can we boost IQ and scholastic achievement?" Harvard Educational Rev. 39:1-123.
KAPLAN, S. (1956) "Social engineers as saviours: effects of World War One on some American liberals." J. of the History of Ideas 17:347.
KARIER, C.J. (1975) Shaping the American Educational State: 1900 to the Present. New York: Free Press.
______VIOLAS, P., and SPRING, J. (1973) Roots of Crisis. Chicago: Rand-McNally.
KATZ, M.B. (1968) The Irony of Early School Reform: Educational Innovation in Mid-19th Century Massachusetts. Cambridge, Mass.: Harvard Univ. Press.
KOLKO, G. (1963) The Triumph of Conservatism. Chicago: Quadrangle Books.
LADD, E.C. and S.M. LIPSET (1975) The Divided Academy: Professors and Politics. New York: McGraw-Hill.
LEVITAN, S. (1969) The Great Society's Poor Law: A New Approach to Poverty. Baltimore: Johns Hopkins Univ. Press.
LEWONTIN, R. (1970) "Race and intelligence." Bull. of Atomic Scientists 26: 2-8.
LIPSET, S.M. and A. BASU (1976) "The roles of the intellectual and political roles," pp. 111-150 in A. Gella (ed.), The Intelligentsia and the Intellectuals. London and Beverly Hills: Sage.
MacKENZIE, D. (1978) "Statistical theory and social interests: a case study." Social Studies of Sci. 8: 35-83.
MANNHEIM, K. (1953a) "Conservative thought," pp. 74-164 in Essays in Sociology and Social Psychology. London: Routledge & Kegan Paul.
______(1953b) "The history of the concept of the state as an organism," pp. 165-182 in Essays in Sociology and Social Psychology. London: Routledge & Kegan Paul.
MARKS, R. (1975) "Race and immigration: the politics of intelligence testers," pp. 316-342 in C. J. Karier (ed.) Shaping the American Educational State: 1900 to the Present. New York: Free Press.
______(1972) "Testers, trackers and trustees: the ideology of the intelligence testing movement in America 1900-1954. Ph.D. dissertation, University of Illinois.
MARRIS, P. and M. REIN (1967) Dilemmas of Social Reform. London: Routledge & Kegan Paul.
MERANTO, P. (1967) The Politics of Federal Aid to Education. New York: Syracuse Univ. Press.
MILLER, S.M. (1978) "Look back in apathy." New Society (22 February): 426-427.
MILLS, C.W. (1966) The Sociology of Pragmatism. New York: Oxford Univ. Press.

MORTON, D.C. and D.R. WATSON (1971) "Compensatory education and contemporary liberalism in the United States: a sociological view." International Rev. of Education 71: 289-308.

MOYNIHAN, D.P. (1969) Maximum Feasible Misunderstanding. New York: Free Press.

_____ [ed.] (1968) On Understanding Poverty: Perspectives from the Social Sciences. New York: Basic Books.

MYRDAL, G. (1944) An American Dilemma: The Negro Problem and Modern Democracy. New York: Harper & Row.

NISBET, R. (1966) The Sociological Tradition. London: Heinemann.

NOBLE, D. (1959) The Paradox of Progressive Thought. Minneapolis: Univ. of Minnesota Press.

NOWOTNY, H. (1975) "Controversies in science: remarks on the different modes of production of knowledge and their use." Zeitschrift für Soziologie 4: 34-45.

PASTORE, N. (1949) The Nature-Nurture Controversy. New York: King's Crown Press.

PIVEN, F.F. and R.A. CLOWARD (1972) Regulating the Poor. London: Tavistock.

ROGIN, M.P. (1967) The Intellectuals and McCarthy. Cambridge, Mass..: M.I.T. Press.

ROSE, S.M. (1972) Betrayal of the Poor: The Transformation of Community Action. Cambridge, Mass.: Schenkman.

SHAPIN, S. (1978) "The politics of observation: cerebral anatomy and social interests in the Edinburgh phrenology disputes," in R. Wallis (ed.), On the Margins of Science. Sociological Review Monographs.

SHUEY, A.M. (1966) The Testing of Negro Intelligence. New York: Social Science Press.

_____ (1958) The Testing of Negro Intelligence. Lynchburg, Va.: Randoph Macon Women's College.

URBACH, P. (1974a, 1974b) "Progress and degeneration in the I.Q. debate." British J. for the Philosophy of Sci. 25: 99-135, 235-259.

WEINSTEIN, J. (1968) The Corporate Ideal in the Liberal State: 1900-1918. Boston: Beacon Press.

WIEBE, R. (1975) The Segmented Society. New York: Oxford Univ. Press.

_____ (1967) The Search for Order. New York: Hill & Wang.

WILLIAMS, R. (1958) Culture and Society. London: Catto and Windus.

YETTE, S.F. (1971) The Choice. New York: Berkley Medallion.

ABOUT THE AUTHORS

BARRY BARNES has lectured in the sociology of science at the Science Studies Unit, Edinburgh University since 1968, having been trained in the natural sciences at Cambridge University and in sociology at Essex University. His main interest is in the sociology of knowledge and culture, and, for a number of years, he has been studying natural science as a typical form of culture, amenable to the normal methods of study of sociology and anthropology. He edited the Penguin anthology of the *Sociology of Science* (1972). Among his publications are books on *Scientific Knowledge and Sociological Theory* (Routledge and Kegan Paul, 1974) and *Interests and the Growth of Knowledge* (Routledge and Kegan Paul, 1977).

ROGER COOTER is presently a Visiting Fellow at the Calgary Institute for the Humanities, University of Calgary, Alberta, Canada. He received his Ph.D. in 1978 from Churchill College, Cambridge, where he completed a dissertation on the ideological significances of popular science in early 19th-century class formations. He has published a number of articles on these subjects and is currently engaged in a book on popular phrenology in the early 19th century.

JOHN DEAN is presently completing a Ph.D. on the recent history of plant taxonomy at the Science Studies Unit, Edinburgh University. He graduated from Edinburgh University in 1975 and worked for short periods in herbaria at Kew and in Edinburgh. His present work combines interests in the history of biology and the sociology of knowledge.

JONATHAN HARWOOD has taught sociology of science in the Department of Liberal Studies in Science, Manchester University, since 1974. He took a Ph.D. in molecular biology at Harvard University in 1970, then studied sociology at Bristol University, and came in 1973 to the Science Studies Unit, Edinburgh University, to begin a study of contemporary controversy over race differences in intelligence. He has published a number of papers on this and related subjects. His research interests lie in the sociology of knowledge and the history of 20th-century biology.

CHRISTOPHER LAWRENCE is the historian to the Wellcome Medical Museum, London. After graduating in medicine at the University of Birmingham in 1970, he worked in general medical practice in the Shetland Islands. In 1973 he read for an M.Sc. in the history of science at Imperial College, London. Since then he has been conducting research on concepts of disease in 18th-century Scotland. He is currently investigating the technology of 19th-century medicine.

DONALD MACKENZIE has lectured in the Department of Sociology, Edinburgh University since 1975. After graduating in applied mathematics at Edinburgh in 1972, he was a research student in the Science Studies Unit there, completing a Ph.D in 1977 on the development of statistical theory in Britain and its relations to political movements and social change. He has published a number of papers on this subject and a book based on his dissertation is currently being prepared for publication by Edinburgh University Press.

ROY PORTER is University Assistant Lecturer and Director of Studies in History at Churchill College, Cambridge, where he completed his Ph.D. in the history of geology in 1974. He has published a number of articles on the history of geology and a book, *The Making of Geology: Earth Science in Britain 1660-1815* (Cambridge University Press, 1977). He is co-editor of *Images of the Earth* (British Society for the History of Science Monographs, 1979). Currently working on a social history of 18th-century Britain and a book on the English Enlightenment, he hopes for the longer term to make a study of the 18th-century British scientific community.

JOAN L. RICHARDS is a graduate student at Harvard University, studying 19th-century history, particularly mathematics, in the History of Science Department. She is in the final stages of a dissertation entitled "William Kingdon Clifford and Non-Euclidean Geometry: A Perspective on the Reception of a Mathematical Theory."

STEVEN SHAPIN is Lecturer in the Social History of Science at the Science Studies Unit, Edinburgh University. He was trained in biology and genetics at Reed College, Oregon, and the University of Wisconsin, and received a Ph.D. in the history and sociology of science at the University of Pennsylvania in 1971. He has published papers on the social history of British science in the 18th and 19th centuries, and has been especially interested in Scottish science and sociological approaches to scientific knowledge. He is presently at work on a book on science and social change in Edinburgh, 1780-1850.

BRIAN WYNNE lectures on sociology of science and technology in the School of Independent Studies, Lancaster University. He obtained a degree in natural sciences at Cambridge University in 1968, followed by a Ph.D. in materials science from Cambridge in 1971. Following this he worked for three years at the Science Studies Unit,